Instructor's Guide with Solutions for Introduction to the Practice of Statistics

SECOND EDITION

George P. McCabe

Linda Doyle McCabe

W. H. Freeman and Company
New York

Cover illustration by Salem Krieger

Major funding for the *Against All Odds: Inside Statistics* telecourse and for the television series has been provided by the Annenberg/CBP Project.

Copyright © 1993 by Consortium for Mathematics and Its Applications (COMAP), Inc. No part of this book may be reproduced by any mechanical, photographic, or electronic process, or in the form of a phonographic recording, nor may it be stored in a retrieval system, transmitted, or otherwise copied for public or private use, without written permission from the publisher.

Printed in the United States of America.

ISBN 0-7167-2450-2

Seventh printing 1996, RRD

Contents

	Data Disk Directory	vii
1	TO THE INSTRUCTOR	1
	1.1 Philosophy and Goals	1
	1.2 Calculations and Computers	2
	1.3 Course Design	4
	1.4 General Comments	10
	1.5 Comments on the Second Edition	12
	1.6 Teaching Suggestions	13
	1.7 Use of the Videos from *Against All Odds*	15
2	COMMENTS ON CHAPTERS	18
3	SAMPLE EXAMS	41
	3.1 SAMPLE EXAMINATION 1	41
	3.2 SAMPLE EXAMINATION 2	50
	3.3 SAMPLE FINAL EXAM	57
4	SOLUTIONS TO EXERCISES	64
	4.1 CHAPTER 1	64
	4.2 CHAPTER 2	106
	4.3 CHAPTER 3	171
	4.4 CHAPTER 4	182
	4.5 CHAPTER 5	200
	4.6 CHAPTER 6	216
	4.7 CHAPTER 7	226
	4.8 CHAPTER 8	248
	4.9 CHAPTER 9	279
	4.10 CHAPTER 10	320

Data Disk Directory

Data is keyed to textbook page.

CHAPTER 1
Example 1.1, page 3
Example 1.5, page 9
Exercise 1.11, page 22
Exercise 1.12, page 22
Exercise 1.17, page 25
Exercise 1.18, page 26
Exercise 1.19, page 26
Exercise 1.20, page 27
Exercise 1.22, page 27
Exercise 1.23, page 28
Exercise 1.25, page 28
Exercise 1.28, page 29
Exercise 1.32, page 31
Exercise 1.33, page 32
Exercise 1.35, page 33
Exercise 1.37, page 34
Exercise 1.39, page 35
Exercise 1.39, page 35,
  Table 1.4, page 36
Exercise 1.41, page 53
Exercise 1.43, page 53
Exercise 1.48, page 53
Exercise 1.49, page 54
Exercise 1.56, page 55
Exercise 1.94, page 83
Exercise 1.108, page 88
Exercise 1.109, page 88
Exercise 1.118, page 91
Exercise 1.119, page 91

CHAPTER 2
Example 2.1, page 98,
  Table 2.1, page 99
Exercise 2.4, page 110
Exercise 2.6, page 111
Exercise 2.8, page 113
Exercise 2.10, page 114
Exercise 2.12, page 115
Exercise 2.13, page 115
Exercise 2.14, page 116
Exercise 2.15, page 116
Example 2.6, Table 2.3, page 118
Example 2.11, page 129
Example 2.21, Table 2.4, page 132
Exercise 2.12, page 137
Exercise 2.30, page 141
Exercise 2.31, page 141
Exercise 2.32, page 142
Exercise 2.35, page 146
Example 2.15, Table 2.5, page 153
Exercise 2.40, page 175
Exercise 2.41, page 158
Exercise 2.42, page 158
Exercise 2.43, page 159
Exercise 2.44, page 159
Exercise 2.46, page 160
Exercise 2.52, page 178
Exercise 2.65, page 182
Example 2.23, page 183
  Table 2.6, page 184
Exercise 2.71, page 193
Exercise 2.72, page 193
Exercise 2.76, page 194
Example 2.26, Table 2.7, page 198
Exercise 2.93, page 207
Exercise 2.94, page 208
Exercise 2.96, page 210
Exercise 2.97, page 210
Exercise 2.99, page 211
Exercise 2.100, Table 2.8,
  page 212
Exercise 2.104, page 213
  Table 2.9, page 214
Exercise 2.105, Table 2.10,
  page 215

CHAPTER 3
Exercise 3.16, page 244
Exercise 3.26, page 246
Example 3.13, page 249
Exercise 3.33, page 254
Exercise 3.35, page 255

Exercise 3.38, page 256
Exercise 3.52, page 270

CHAPTER 4
Exercise 4.72, page 355
Exercise 4.73, page 356

CHAPTER 5
Example 5.20, page 409,
　Table 5.1, page 410
Exercise 5.49, page 416,
　Table 5.2, page 417
Exercise 5.50, page 417
Exercise 5.51, page 418
Exercise 5.56, page 419
Exercise 5.57, Table 5.3, page 420
Exercise 5.58, page 420

CHAPTER 6
Exercise 6.11, page 444
Exercise 6.12, page 444
Exercise 6.31, Page 468
　(Same as 6.11, page 444)
Exercise 6.33, page 369
　(Same as 6.12, page 444)
Exercise 6.41, page 470

CHAPTER 7
Example 7.3, page 507,
　Table 7.1, page 508
Exercise 7.14, page 522
Exercise 7.15, page 522
Exercise 7.18, page 523
Exercise 7.20, page 524
Exercise 7.33, page 528
Example 7.8, Table 7.2, page 534
Example 7.11, page 540
Example 7.12, page 543,
　Table 7.3, page 544
Exercise 7.44, page 552
Exercise 7.45, page 552
Exercise 7.73, page 567
Exercise 7.79, page 569

CHAPTER 8
Exercise 8.29, page 598
Example 8.10, page 601
Example 8.14, page 608
Exercise 8.38, page 621
Exercise 8.39, page 622
Exercise 8.40, page 622
Exercise 8.41, page 622
Exercise 8.42, page 623
Exercise 8.43, page 624
Exercise 8.53, page 627
Exercise 8.54, page 627
Exercise 8.55, page 628
Exercise 8.56, page 628
Exercise 8.65, page 630
Exercise 8.69, page 631
Exercise 8.70, page 632
Exercise 8.71, page 632
Exercise 8.73, page 634

CHAPTER 9
Example 9.7, page 659
Exercise 9.1, page 674
Exercise 9.3, page 675
Exercise 9.7, Table 9.3, page 678
Exercise 9.16, page 681
Exercise 9.22, page 683
Exercise 9.23, page 684

CHAPTER 10
Example 10.6, Table 10.1,
　page 725
Exercise 10.25, page 757
Exercise 10.26, page 758
Exercise 10.55, page 786
Exercise 10.56, page 787

APPENDIX
Cheese, page 792
CSData, page 793
Wood, page 794
Reading, page 795
Majors, page 796

1 TO THE INSTRUCTOR

1.1 Philosophy and Goals

The text is intended to be used in a first course in statistics for students with limited mathematical background. It covers the basic material presented in many courses at this level—data analysis, a little probability and standard statistical methods. The emphasis is on understanding how to use statistics to address real problems. In this regard it differs from most introductions.

Many students view statistics as a collection of formulas that give correct results if the computations are performed accurately. The competence of the student is measured by the complexity of the formula that can be computed correctly. Unfortunately this view is often reinforced by the way statistics is traditionally taught.

With the advent of modern computers, complexity of computational formulas is no longer an issue. As a result we are now free to concentrate on the practice of statistics as a tool for learning about the real world. Consider, for example, a pre-election poll taken for a major news organization. The results are reported as percentages with a margin of error. Typically stratification is used and perhaps even cluster sampling. The formulas for the standard error are complex. However, important issues for understanding and using the results relate to how the question is phrased and how the data were obtained. If students are able to analyze and interpret data taken by a simple random sample, then the transfer of the basic ideas to the more complex situation is easy. The meaning of a confidence interval, for example, does not depend upon the formula used to calculate the standard error.

In using the text it is clear that students at this level enjoy discussing ideas related to such topics as the numbers of home runs hit by Ruth and Maris (Example 1.4), the use of gastric freezing as a treatment for ulcers (a treatment indistinguishable from a placebo—Example 3.5) and the prediction of college grade point average by SAT scores and high school grades (Section 2 of Chapter 9). They do not enjoy spending an hour trying to use a complicated formula to obtain a numerically correct result. The text reflects this philosophy. To be used effectively, it is necessary for the instructor to project this view.

Among mathematicians and statisticians who teach introductory statistics courses, there is a tendency to view students who are not skillful in mathematics as unintelligent. There is a large group of students who seem to turn off their common sense as soon as they enter our classrooms. Experience with this text suggests that these students can be persuaded to turn on their common sense and to think quite deeply about real problems that involve statistics. The key is for them to realize that statistical methods are tools for answering interesting questions rather that a collection of formulas.

There are very few people who find formulas exciting to look at or use. However, it is very exciting to realize that a control chart can be used to improve the quality of the products made by industry and that the statistical analysis of data from a properly designed experiment can be used to compare the effectiveness of a new pharmaceutical with a standard treatment. The formulas and calculations we use are thus interesting (and worth learning) because they allow us to answer real questions.

Upon completion of a course based on this text, students should be able to think critically about data, to use graphical and numerical summaries, to apply standard statistical inference procedures and to draw conclusions from such analyses. Detailed comments regarding particular chapters are given in the section COMMENTS ON CHAPTERS of this guide. For students wanting to pursue further study of statistics, the basics provided by this material should be adequate for further courses on topics such as experimental design, regression, sample surveys and quality control.

1.2 Calculations and Computers

Clearly the practice of statistics requires a fair amount of numerical calculations. Furthermore, some calculation helps facilitate the understanding of a particular technique or method. On the other hand, particularly for students at this level, it is very easy for the computational aspects of a problem to interfere with a full understanding of the concepts.

Consider, for example, the standard deviation. It is much more important that a student be able to apply the 68-95-99.7 rule for normal distributions described in Section 3 of Chapter 1 than to be able to quickly calculate the standard deviation for a sample of size 50 using the computing formula given in equation 1.3. The defining formula given in equation 1.2 is more useful for understanding but is rather cumbersome for calculation, particularly with real data. It is difficult to imagine any meaningful understanding coming from the use of equation 1.3. Inexpensive calculators with functions for $\bar{x}$

1.2 Calculations and Computers

and s can be used effectively for calculations.

The text is designed to be used both in courses where students have access to computing facilities and where they do not. As computing continues to become cheaper and more readily available, it is expected that use of computers in courses at this level will increase. Most of the examples used in the text involve computations that can readily be performed with a calculator rather than a computer. Each chapter contains exercises with small amounts of data or with the results of some computations given that are suitable for students without a computer. However, students with access to computing can be suitably more challenged by the exercises with real data requiring use of a statistical package. The COMPUTER EXERCISES at the end of each chapter are included for this purpose.

Although the use of a computer for a course based on this text is not required, most people who *do* statistics today use a computer. Therefore, any serious attempt at explaining the practice of statistics must recognize this fact. In the early chapters of the text, examples of MINITAB commands and output are presented. In the last three chapters SAS outputs are given. These two packages were chosen on the basis of their widespread popularity and availability. Both are of high quality. The choice of these packages is in no way intended to suggest that other packages not be used or even that both of these be used in the same course. The text does not attempt to show students how to use the computer. Supplementary material is clearly needed for this purpose. A computer package is a statistical tool—a means for performing arithmetic and generating plots. Many other packages could easily be used in a course based on this text.

For students who are expected to take only one course in statistics and are unlikely to encounter large amounts of data, something like MINITAB is most appropriate. Such packages are becoming popular on personal computers. Unfortunately the number of these packages is increasing rapidly and some are of rather poor quality. See the recent article in the *American Statistician* by Gerald Dallal titled "Statistical microcomputing—like it is," (August 1988, p. 212-216).

On the other hand, students who will be taking additional advanced courses and are likely to use more complicated procedures are better prepared for this by learning to use a more elaborate package. SAS, BMDP and SPSS are the major competitors in this end of the market. All three are products of high quality and each has its own strong points.

1.3 Course Design

In designing a course based on this text, some selection of topics must be made. Very few groups of students are capable of assimilating all of the material in detail in a single semester or quarter. The amount of time needed to cover any given topic depends to a large extent on the mathematical sophistication of the students.

Consider, for example, the second section of Chapter 2 on least–squares regression. Here the idea of using a line to express something about the relationship between two variables is introduced. For many students translating an equation into a line on a plot is a non-trivial task. They may have encountered the idea in a course on algebra but applying it to a real problem is a different matter. Such students need to have these ideas carefully explained and should work through exercises such as 2.16 to 2.20. For students with a better mathematical background, this material is quite easy and can be treated briefly. A key statistical concept in this section is the idea of a residual. The theme DATA = FIT + RESIDUAL recurs many times in later chapters. If the students do not grasp the idea of FIT, expressed as a straight line in this section, then they will be unable to learn the meaning of a residual.

The text is an elementary but serious introduction to modern statistics for general college audiences. This means that the material is presented in a way that it can be learned by students who are not particularly skillful in mathematics. As the example above points out, the amount of time required for learning and hence, the amount of material that can be covered, depends upon the skill level of the students in a particular course. One further consideration is important in the selection of material. The text has been used for beginning graduate students in non-quantitative disciplines who will be analyzing data for their research. Many of these students will take another course in statistics but some will not. Such students have a real need to be exposed to the methods presented in the last three chapters—analysis of count data, regression and analysis of variance. Accomplishing this for students with moderate mathematics skills requires skipping or treating lightly many of the optional sections in addition to a great deal of work on the part of the students. Experience has shown that this type of student is highly motivated and willing to work quite hard. The practical relevance of the examples and exercises is an important factor for sustaining the motivation

1.3 Course Design

and drive that these students bring with them to the first lecture.

On the other hand, a course for undergraduates who will be consumers rather than producers of statistically-based research should emphasize the unstarred portions of Chapters 1 to 8. Any of the material in 8.3, 9.1, or 10.1 can be covered at the option of the instructor. But we encourage allotting adequate time to the earlier chapters even if the course ends at 8.2 or earlier.

The following are some sample outlines for the coverage of topics. They are presented as a aid for the instructor designing a course rather that a strict set of rules for how it should be done. Each outline gives the number of weeks for the course, the mathematical skill level of the students and the desired breadth of coverage.

OUTLINE I: 15 Weeks, Low or moderate skills, Narrow coverage

Week	Material to be Covered
1	Chapter 1, Sections 1 and 2
2	Chapter 1, Section 3; Chapter 2, Sections 1, start Section 2
3	Chapter 2, Complete Section 2, Section 4 (skip Section 2.3)
4	Chapter 2, Sections 5 and 6
5	Review Chapters 1 to 2; Exam I on Chapters 1 to 2
6	Chapter 3, Sections 1, 2, 3 and 4
7	Chapter 4, Sections 1 and 2
8	Chapter 4, Section 3 (skip 4.4); Chapter 5, Section 1
9	Chapter 5, Section 2 (skip 5.3); Review Chapters 3, 4 and 5
10	Exam II on Chapters 3, 4 and 5
11	Chapter 6, Section 1 and first two parts of Section 2
12	Chapter 7, last part of Section 2 and first two parts of Section 3 (skip last two parts of Section 3, power and inference as a decision)
13	Chapter 7, Sections 1 and 2 (skip Section 3)
14	Chapter 8, Sections 1 and 2
15	Review
Comprehensive Final Exam	

For students with moderate mathematics skills, a little more depth can be taught for each topic and some of the more challenging Chapter Exercises can be assigned or discussed in class.

OUTLINE II: 15 Weeks, High skills, Narrow coverage	
Week	Material to be Covered
1	Chapter 1, Sections 1, 2 and 3
2	Chapter 2, Section 1 and start Section 2
3	Chapter 2, Complete Section 2, Section 4
4	Chapter 2, Sections 5 and 6; Chapter 3, Section 1
5	Chapter 3, Sections 2, 3 and 4
6	Review Chapters 1 to 3; Exam I on Chapters 1 to 3
7	Chapter 4, Sections 1, 2 and 3
8	Chapter 4, Section 4; Chapter 5, Section 1
9	Chapter 5, Sections 2 and 3; Review Chapters 4 and 5
10	Exam II on Chapters 4 and 5
11	Chapter 6, Sections 1 and 2
12	Chapter 6, Section 3
13	Chapter 7, Sections 1, 2 and 3
14	Chapter 8, Sections 1, 2 and 3
15	Chapter 9, Section 1 (skip optional parts) ; Review
	Comprehensive Final Exam

Outline II covers all of the topics in Chapters 1 through 8 and part of Chapter 9. A course based on this outline would give students a solid introduction to the basic ideas of data analysis and statistics while omitting much of the linear regression chapter and all of analysis of variance.

1.3 Course Design

OUTLINE III: 15 Weeks, Moderate or High skills, Broad coverage

Week	Material to be Covered
1	Chapter 1, Sections 1, 2 and 3
2	Chapter 2, Section 1 and 2 (skip Section 2.3)
3	Chapter 2, Sections 4, 5 and 6
4	Chapter 3, Sections 1, 2, 3 and 4; Review Chapters 1 to 3
5	Exam I on Chapters 1 to 3; Chapter 4, Section 1
6	Chapter 4, Sections 2 and 3 (skip 4.4)
7	Chapter 5, Sections 1 and 2 (skip 5.3)
8	Chapter 6, Sections 1 and 2
9	Chapter 6, Section 3; Chapter 7, Section 1
10	Chapter 7, Sections 2 and 3; Exam on Chapters 4 to 7
11	Chapter 8, Sections 1, 2 and 3
12	Chapter 9, Section 1
13	Chapter 9, Section 2
14	Chapter 10, Section 1
15	Chapter 10, Section 2
Comprehensive Final Exam	

Outline III is designed to reach the important applied topics presented in the last three chapters. To accomplish this goal it is necessary to skip topics that are not essential in the earlier chapters. In particular the coverage of probability is reduced to the minimum necessary for understanding these inferential procedures. In contrast to Outlines I and II, there is no review time specified in this outline. By eliminating the optional sections 9.2 and or 10.2, time for this purpose can be incorporated into the course.

OUTLINE IV: 10 Weeks, Low or Moderate skills, Narrow coverage

Week	Material to be Covered
1	Chapter 1, Sections 1, 2 and 3
2	Chapter 2, Sections 1 and 2 (skip Section 2.3)
3	Chapter 2, Sections 4, 5 and 6
4	Chapter 3, Sections 1, 2, 3 and 4; Review Chapters 1 to 3
5	Exam on Chapters 1 to 3; Chapter 4, Section 1
6	Chapter 4, Sections 2 and 3 (skip 4.4)
7	Chapter 5, Sections 1 and 2 (skip 5.3)
8	Chapter 6, Sections 1 and 2
9	Chapter 6, Section 3; Chapter 7, Section 1
10	Chapter 7, Sections 2 and 3
	Comprehensive Final Exam

For a quarter course of 10 weeks the coverage is necessarily narrow and the options available are somewhat restricted. Outline IV is designed to reach the basic inferential material given in Chapter 7 for normal problems with unknown standard deviations. The topics included are those necessary to reach this point. The depth to which each topic is explored can be varied depending upon the skill level of the students.

OUTLINE V: 10 Weeks, High skills, Narrow coverage

Week	Material to be Covered
1	Chapter 1, Sections 1, 2 and 3
2	Chapter 2, Sections 1, 2 and 3
3	Chapter 2, Sections 4, 5 and 6
4	Chapter 3, Sections 1, 2, 3 and 4
5	Exam on Chapters 1 to 3; Chapter 4, Sections 1 and 2
6	Chapter 4, Sections 3 and 4
7	Chapter 5, Sections 1, 2 and 3
8	Chapter 6, Sections 1 and 2
9	Chapter 6, Section 3; Chapter 8, Section 1
10	Chapter 7, Sections 2 and 3
	Comprehensive Final Exam

1.3 Course Design

For those using a quarter system, a two course sequence based on the text provides the opportunity for broad coverage of the topics with variation on depth depending upon the skill level of the students. The following two outlines are designed for such a sequence.

OUTLINE VI: 10 Weeks, First Course of a Two Course Sequence

Week	Material to be Covered
1	Chapter 1, Sections 1, 2 and 3
2	Chapter 2, Sections 1 and 2
3	Chapter 2, Sections 3 and 4
4	Chapter 2, Sections 5 and 6; Review Chapters 1 to 2
5	Exam on Chapters 1 to 2; Chapter 3, Sections 1 and 2
6	Chapter 3, Sections 3 and 4; Chapter 4, Section 1
7	Chapter 4, Sections 2 and 3
8	Chapter 4, Section 4; Chapter 5, Section 1
9	Chapter 5, Sections 2 and 3
10	Review
	Comprehensive Final Exam

OUTLINE VII: 10 Weeks, Second Course of a Two Course Sequence

Week	Material to be Covered
1	Chapter 6, Sections 1 and 2
2	Chapter 6, Section 3; Chapter 7, Section 1
3	Chapter 7, Sections 2 and 3
4	Review Chapters 6 and 7; Exam on Chapters 6 and 7
5	Chapter 8, Sections 1 and 2
6	Chapter 8, Section 3; Chapter 10, start Section 1
7	Chapter 9, Complete Section 1, Section 2
8	Chapter 10, Section 1
9	Chapter 10, Section 2
10	Review
	Comprehensive Final Exam

1.4 General Comments

When students complete a course based on this text they should be able to analyze real data and draw conclusions. Since the emphasis is on *doing* statistics rather than talking about statistics, much of the class time should be spent discussing how different analyses tell us something about data. The methodologies presented are all based upon sound statistical theory. Details of how the theory leads to particular formulas are neither interesting to the students nor are they essential to the central purpose of learning to do statistics. A minimum amount of time should be spent on examining formulas and procedures in the abstract. Theoretical ideas need to be understood in the context of how they *apply* to the analysis of data.

Consider, for example, the problem of outliers. They are a fact of life for anyone applying statistics to real problems. It is not sufficient to take a strict mathematical view and assume that they do not exist. A procedure based on normal theory may give very misleading conclusions if applied data contaminated by outliers. In the practice of statistics it is important to be able to identify outliers and with an understanding of the area of application, to do something about them. On the other hand, it is rather unimportant to know that a t distribution can be represented at the ratio of a standard normal to the square root of a chi-square.

To teach students how to do statistics, you will spend most of your class time doing statistics. This means discussing data, not just numbers devoid of any real meaning. It means saying something about the context in which the data were collected and the field of application. Instructors in statistics courses typically are very comfortable with the rather elementary mathematical and statistical ideas needed to discuss the methods presented in the text. On the other hand, few are experts in the various substantive fields from which the examples and exercises are drawn. Most of these, however, involve only a basic understanding much of which is presented with the example or exercise.

Using data from real problems provides many opportunities for the students to become actively involved in the class. For example, they can be asked in class to explain what a home run or an at bat is for the foreign students who may never have heard of this term. By doing so, they gain confidence and feel that they have something meaningful to contribute. At a first glance these ideas may appear to be rather simple to explain but expe-

1.4 General Comments

rience has shown that this is often not so and attempts at explanation can lead to interesting discussions. For example, not all home runs are hit out of the park and perhaps some would advocate using a different definition of at bats than the standard one.

Incomplete knowledge of all of the fields covered by the examples and exercises can be turned into a very effective teaching tool. Students have diverse backgrounds and often know a great deal more than we think. It is a very pleasant experience to leave a class session having learned something from the students. When discussing the corn yield data presented in Example 3.3 one time, I was unable to remember whether or not the fields were irrigated but I did remember that they were in Nebraska. I mentioned this and one young lady in the class immediately raised her hand and explained that if it was corn in Nebraska, then it certainly was irrigated! This sort of experience cannot easily be preplanned in your lecture notes. It occurs naturally in a teaching environment where the students are encouraged to think about data rather than numbers.

There is a tendency today to try to design statistics courses specifically for students in one particular major or general field. This is particularly true for students in business and the social sciences. Even in general courses some students express the opinion that they would prefer examples and homework based on problems coming exclusively from their own field of interest. If we are attempting to educate rather than to simply train students, this view is very short-sighted. Ideas learned well in one context are easily translated to others. For example, Roger Maris' 61 home runs do have something in common with the agricultural production in the drought year of 1988. One of the interesting things about statistics as a field is that the same sorts of fundamental ideas occur in many diverse settings. By seeing these in different settings learning is facilitated. Furthermore, many fields are not as homogeneous as they might appear at a first glance. Often, interesting problems require the assimilation of material from diverse areas. This is particularly true of business. In summary, do not apologize to your students because all of the examples are not from their field. Explain to them that breadth is part of education.

There are a sufficient number of Examples in the text for some of these to be used in class lectures. Sometimes it is useful to build a lecture around one or a sequence of Exercises. Students will sometimes bring in examples that they have found in newspapers or elsewhere. Using such material adds a nice personal flavor to the course and encourages class participation.

If computers are used, it is a good idea to run some of the Examples or Exercises on the system you are using. Copies of output can be distributed to the students and serve as the basis for some lectures. Outputs can also be enlarged and made into transparencies.

The practice of statistics is learned by doing statistics rather than by reading about it. Many students have difficulty with this idea since it is not part of the learning strategy employed in many of their other courses. To master the material in the text the students need to work solutions to the exercises—many of them. A lot of homework given frequently is recommended. A large number of exercises require little or no computation. These can be used to learn the basic ideas with being concerned about obtaining the correct number from a formula. Note that solutions are given for the odd-numbered Exercises in the back of the text. For each of these there is generally an even-numbered exercise with no solution having similar characteristics.

1.5 Comments on the Second Edition

In the Second Edition, we have made changes designed to make the book more accessible to students and more teachable for instructors. There are numerous small improvements in the writing, simpler notation where possible and several interesting new real data sets.

Most teachers were treating the topics in Chapter 2 very briefly. Therefore, we have integrated this material with the appropriate topics in other chapters. Plots against time appears in Chapter 1 and exponential growth is an optional section in the new Chapter 2. Statistical control is discussed in Chapter 5. The new Chapter 2 contains the material from the old Chapter 3 with the above modifications. This change has resulted in a renumbering of old Chapters 4 through 11; they are now 3 through 10.

The probability material in Chapters 4 and 5 has been rewritten to allow for easier comprehension by the students. Bayes's rule appears in an optional section.

We have rewritten the presentation of the chi-square test for two-way tables in Section 8.3 giving a more direct path to the main results. The same is true for inference in simple linear regression in Section 9.1.

A new section at the end of each chapter gives a collection of exercises designed to be solved with a computer. A data disk is available in several formats containing all large and several moderate-size data sets. A completely new Minitab Guide is available.

1.6 Teaching Suggestions

Statistics has a well-earned reputation as the dullest of subjects. We teachers have the responsibility of overcoming that preconception by demonstrating to students that our subject is both intellectually stimulating and useful. We believe that almost any statistics course can be improved by including more data and more emphasis on reasoning, at the expense of fewer recipes and less theory. "Data" means not just numbers, but problems set in a context that require a conclusion or discussion rather than just a calculation or graph. In *Introduction to the Practice of Statistics* we have tried to emphasize data and reasoning, limited of course by our own ability and by the need to present an exposition accessible to beginning students. We hope the text will at least not stand in the way of teachers who want their students to come away with more than a list of recipes. Here are some suggestions from our experience.

Involve the class. The variable style mixing blackboard work with direct address (get out from behind the lectern) helps keep students awake. Even better is discussion with the class. Use some exercises as bases for discussion, assigning them "for discussion" in the next class rather than requiring written answers. Ask a student to present an analysis and then give a non-technical conclusion, as if reporting to a boss. Then ask the other students to respond with questions or comments.

Leading discussions is interesting but difficult. You must resist the temptation to leap in with the "correct" view. Try rather to guide the class by questions and to get other students to offer alternatives to weak answers. Be patient when asking questions on the spot. An educational psychologist of our acquaintance notes that teachers rarely wait long enough for students to assimilate a question and produce a response. He suggests 30 seconds, which seems an eternity while passing in silence but does produce more response in the end. Above all, don't put down a student who is incorrect or confused. Students notice even the silent disappointment that contrasts so clearly with your response to an intelligent answer. Try to give a positive response to every student who is brave enough to speak up – they are helping you. When you know the class better, you can direct easier questions at the weaker students, thus preventing monopoly by the bright and aggressive. Remember that you are building confidence, not simply conveying information. Discussion isn't easy for teachers oriented toward problem-solution or theorem-proof presentation. But the attempt is essential if your students are afraid of statistics and perceive it as remote from their experience.

Lecturing. If you have reasonably small classes, don't lecture all the time. After putting the main points on the board as an outline, guide a discussion of the day's topic. Insist that students do the reading in advance so that they can participate, and make it clear that class participation is part of their grade. Discussion will sometimes take the form of getting straight what the text means, and often of applying the ideas in examples that you pose for the class.

Many of us, however, must lecture. One course we teach has lectures to 200 students twice a week, with recitation sections of 30 students led by teaching assistants on a third day. Education researchers insist that lectures are a second-rate way of helping students learn. But that is not to say that they are without value. Some students are better able to follow a lecture than to read a text. We must do our best at a less-than-ideal form. Unless your students are unusually skilled readers, your lectures must help them see the main path of ideas and must underline the main points. When you lecture,

- Outline the basic points on the blackboard or overhead. Don't hesitate to repeat the text, but be schematic – students love numbered heads and subheads in their notes.

- Give added examples both to motivate and to instruct.

Use supplements. Readings from *Statistics: A Guide to the Unknown* (Third edition, Wadsworth and Brooks/Cole 1989, edited by Judith Tanur, et al.) illustrate the richness of statistics. Even better are excerpts from the Annenberg/Corporation For Public Broadcasting video series *Against All Odds: Inside Statistics* See the next section for details on how these videos can be used in your course.

1.7 Use of the Videos from *Against All Odds*

The Annenberg/Corporation For Public Broadcasting video series *Against All Odds: Inside Statistics*, for which David Moore was the content developer, is an excellent source of supplementary material. This series of 26 half-hour programs is available on VHS videotape at an outrageously low price ($350 as we write: call 1–800–LEARNER for information or to order). It is a telecourse that presents the content of a first course in statistical methods, originally intended for teaching of distant learners. *Introduction to the Practice of Statistics* was written in conjunction with *Against All Odds*, so the telecourse follows the text quite closely.

Television is not an adequate substitute for a live teacher. We don't recommend showing entire programs from *Against All Odds* in on-campus instruction. But video has several strengths that make it an ideal supplement to your own teaching. Television can bring real users of statistics and their settings into the classroom. And psychologists find that television communicates emotionally rather than rationally, so that it is a vehicle for changing attitudes. We suggest showing a number of the short documentary segments that were filmed on location. Not only will students enjoy the break, but seeing a variety of people using statistics in a variety of settings is much more effective in changing attitudes than simply saying that statistics is useful. Consider using some of these video documentaries as starters for discussion by asking students to apply the concepts and tools they are learning to the setting in the video. In several cases, data from the video appear in the text. You may also decide to show some of the computer graphics used for teaching in the videos. A good example is the graphic in Program 19 that illustrates the behavior of confidence intervals in repeated sampling. Here are some suggestions for effective short documentaries from *Against All Odds*. The parenthesized notes tie the stories to sections of the text.

- *What is Statistics?*, the collage of excerpts from later stories that makes up the first half of Program 1. You can order an edited 14-minute version under this title from the American Statistical Association. We recommend this even if you use no other video material. (Introduction.)

- *Lightning research* from Program 2; histograms and how to look at distributions in the setting of a study of lightning in Colorado. (Section 1.1.)

- *Calories in hot dogs* from Program 3; the five-number summary and box plots compare beef, meat, and poultry hot dogs. (Section 1.2; see Table 1.4 on page 36 of IPS for the data.)

- *The Boston Beanstalk Club* from Program 4. This social club for tall people leads to discussion of the 68–95–99.7 rule for normal distributions. (Section 1.3.)

- *Saving the manatees* from Program 8, in which the linear relation between power boats registered in Florida and manatees killed by boats introduces least squares fit. (See Exercise 9.1 on page 674 of IPS for updated data and a Minitab regression output. You can use the story to accompany Section 2.2 or Section 9.1.)

- *The Minnesota twin study* in Program 9 is an important study of the correlation in physical and mental measures in twins raised apart. (Section 2.4. Note that for twin studies the intraclass correlation is used because the twins in a pair are interchangeable. The interpretation is the same for our purposes as that of the usual correlation.)

- *Smoking and health* in Program 11 takes a historical look at the search for evidence of causation when experiments can't be done. (Section 2.6.)

- *The Physicians' Health Study* from Program 12 is a major clinical trial (aspirin and heart attacks) that introduces design of experiments. To start a discussion on the ethics of experiments with human subjects, show this excerpt with the following story on a social policy experiment to study the effects of police response to domestic violence calls. Dr. Hennekens, the director of the Physicians' Health Study, comments on the ethical standards that constrain clinical trials; in contrast, the domestic violence study does not even ask consent of its subjects. Some medical statisticians viewing this program have called the domestic violence study clearly unethical. But, as a statistician involved in analyzing domestic violence data said, "These people elected themselves for the study by committing an act that justifies arrest." (Section 3.2.)

- *Sampling at Frito-Lay* from Program 13 illustrates the many uses of sampling in the context of making and selling potato chips. This is one

1.7 Use of the Videos from *Against All Odds*

of the visually most interesting stories in *Against All Odds*. (Section 3.3.)

- *Binomial examples* from Program 17. Show the brief excerpts that present the "hot hands in basketball" issue, heredity (sickle-cell anemia), and a large quincunx in a science museum. Ask your students: What is a "success" in each case? Are counts of successes in these settings really binomially distributed? Which aspects of the binomial setting are clearly present and which are debatable? Do you think that lots of data could resolve the debatable points? (Section 5.1.)

- *Control charts at Frito-Lay* from Program 18 is good if you did not show the sampling excerpt from Program 13 shot at the same location. (Section 5.3.)

- *Battery lifetimes* from Program 19, with the following animated graphic that illustrates the behavior of confidence intervals in repeated sampling. (Section 6.1.)

- *Welfare reform* in Baltimore, from Program 22, is an important comparative study of new versus existing welfare systems that leads to a two-sample comparison of means. (Section 7.2.)

- *The Salem witchcraft trials*, revisited in Program 23, show social and economic differences between accused and accusers via comparison of proportions. (Section 8.2.)

- *Medical practice*: Does the treatment women receive from doctors vary with age? This story in Program 24 produces a two-way table of counts. (Section 8.3.)

- *The Hubble constant* relates velocity to distance among extra-galactic objects and is a key to assessing the age of the expanding universe. A story in Program 25 uses the attempt to estimate the Hubble conference to introduce inference about the slope of a regression line. (Section 9.1.)

2 COMMENTS ON CHAPTERS

Chapter 1 Looking at Data: Distributions

Students taking a first course in statistics often do not know what to expect. Some may view statistics as a field where the major task is to tabulate large collections of numbers accurately. Others have heard that statistics is more like mathematics with a lot of complicated formulas that are difficult to use. Few are expecting a course where they need to use their common sense and to *think*.

The presentation of the material in Chapter 1 sets the tone for the entire course. Statistics is a subject where intelligent judgements are needed and common sense plays a large role. To the extent that you can get the students to think deeply about the material at this stage, you will have succeeded in setting the proper tone. Most of the examples presented in the text can serve as the basis for extensive class discussions. Depending upon the interests of the students, some of these may be more suitable than others. You can also try to find or collect data that the particular students you are teaching can relate to.

In one class based on this text, students were asked to take their pulse on the first day of class. No particular instructions concerning how to take one's pulse were given. The data that resulted were used in discussions that illustrated several major themes of the course. A stemplot was immediately constructed. There was one very large outlier—180 beats per minute. We discussed whether or not this value was reasonable. Several of the physical education students commented that someone who was very out of shape might elevate their pulse to such a high level by running up to the third floor where the course was held. Someone from psychology mentioned that a person who was very anxious about the course might have an elevated pulse. One conclusion drawn was that if you want to measure anything on people, you must pay attention to the circumstances present when the measurement is taken.

Further examination of the data revealed an exceptionally large number of readings that ended with the digit zero. Students in aerobics classes typically take their pulse for 6 seconds and multiple by 10. This led to a discussion of the possibility that different methods were used to obtain the reported pulse rates. Since it was impractical to ask all 50 or students what method each used, a random sample was taken using the class list and the table of random numbers given in the back of the text (Table B). Students had no trouble following the method. Successive pairs of digits were used and numbers outside the range 1 to 50 were discarded. It was not surprising to find a number corresponding to a student who was not present in the class on that day. A short discussion about missing data and the problems associated with statistics in the *real* world followed.

The methods reported by the sampled students were very revealing. Most were what you would expect. Measure for 60 seconds, measure for 30 seconds and multiply by 2, measure for 15 seconds and multiply by 4, and of course, those who had taken an aerobics course. One said that he did not have a watch so the value given was a guess. Another (a very polite foreign student whose command of spoken English was not very good) explained that she had no idea about what was asked and simply reported the value given by another student.

The lesson was quite clear. It is not easy to gather good data. What is measured and how it is measured are the two most important questions to ask when undertaking any statistical analysis. Understanding the issues related to these questions does not require mathematical sophistication or the ability to calculate $\sqrt{1/n_1 + 1/n_2}$ by hand. If the students can be sufficiently motivated at the beginning of the course and understand that many interesting things can be learned by looking at data, they will be willing to put in the effort required to master the skills needed for the computations required later.

Look for sets of data that will be interesting to your students. They are not very difficult to find. If the students are concentrated in a few majors, ask faculty from those departments for reprints of articles using elementary statistical methods. Ask questions about how things are measured and whom or what is measured. Do the items or people measured represent any larger group or do they simply represent themselves? Usually we think of statistics as being used used only for inference to some larger known or imagined population. However, many interesting sets of data do not fall into this category and much about data analysis can be learned by studying them. Government statistics often come from a census or samples so large that the sampling variability is negligible. A good job of describing such data is often not a trivial exercise.

The key concepts in this chapter concern ways of looking at and describing one set of data. We start with graphical displays and then move to numerical summaries. The point of the numerical summary is that it quantifies and expresses compactly, something that can be seen roughly in a graphical display. The effectiveness of a numerical summary is based on how well it calls to mind what appears in the graphical display. Thus viewed, the mean is not a very good numerical summary for a set of data with two distinct clusters of observations. From numerical summaries we proceed to models for data, illustrated by the important case of the normal distributions. The progression from graphical display to numerical measures to a compact mathematical model is one of the strategies that unifies data analysis and makes it more than a collection of clever methods.

1.1 Displaying Distributions

The introductory material on measurement is short but important. What is measured and how it is measured are two fundamental issues that lie at the foundation of any statistical analysis. Many of the judgements made at this level are subjective and require common sense. Make the point that statistics is about data and not just numbers.

Variation is another key concept. Why are all of the observations is a set of data not the same? Good class discussions can be based on sources of variation in particular sets of data.

The first real statistical analysis presented is the stemplot. Stress the idea that we start with a graphical display before proceeding to numerical summaries. We have presented stemplots with the values increasing as you proceed down the stems. Some computer packages put the larger numbers on the top. Similarly, in the text truncation is recommended (this is clearly the easiest method to use if constructing the plot by hand), whereas some programs will round. Keep in mind that the purpose of the plot is to get a quick look at the data and to help in deciding what sorts of numerical summaries are appropriate. Be flexible. Splitting leaves and other sorts of judgements should be based on common sense rather than on adherence to a set of inflexible rules.

Similar comments regarding flexibility apply to histograms. The histogram is presented as a method for displaying a set of data when a stemplot is unsuitable. It also serves an important role pedagogically. In the last section of this chapter, the density curve is introduced as a model that describes the shape of a histogram when the sample size increases and the widths of the intervals decrease.

The major theme of this section is that we should always carefully look at data. In graphical displays we look for an overall pattern and deviations from that pattern. Simon Newcomb was a famous scientist who was measuring well-defined physical quantities. If he had outliers in his data, we should not be surprised to find them in other sets of data that we may encounter. This section introduces the basic strategy for looking at data: first seek an overall pattern, then deviations from the pattern. This strategy will be used in other settings in Chapters 2 and 3.

1.2 Describing Distributions

The mean should be familiar to almost all students. Therefore, the introduction of the summation notation is facilitated by noting that it is a compact way to express something that they already know. Although the idea behind the median is quite intuitive, some students get confused by the manipulations required to find it. Emphasize that it is a numerical measure for something that can be seen in a stemplot or histogram.

Quartiles and the five number summary build on the idea of the median. Again emphasize that they are numerical summaries for something that can be seen graphically. With the boxplot we complete a circle of sorts. Starting with a graphical summary, we calculate numerical summaries and then present these graphically. Note that there are many variations on boxplots. If you use statistical software, the rules for constructing them may differ slightly from those presented in the text. This should not pose any serious problems. In fact, it can be used to illustrate the point that caution is needed in using statistical packages. We need to know how things are computed if we are to interpret the output meaningfully.

No single numerical summary is appropriate for all sets of data. There is a tendency to think that robust or resistant measures always work (although they may be inefficient). The dates of founding for New Zealand wineries given in Exercise 1.100 are an example of real data where the boxplot fails to capture very important aspects of the distribution.

The standard deviation is a fundamental quantity, the meaning of which will be made clear in the last section on the normal distributions. Here, the variance is a necessary intermediate step required for the calculation of the standard deviation. For students who have a great deal of trouble with computations, the calculations can be organized in a table with columns corresponding to x, x^2, $x - \bar{x}$ and $(x - \bar{x})^2$. We recommend that students have a calculator that computes $\bar{x}$ and s from keyed-in data. This inexpensive tool greatly reduces the arithmetical burden for many standard statistical procedures.

Linear functions are used in several places later in the text. Thus, the material on changing the unit of measurement has two purposes. One idea is that when data are transformed in this way, nothing fundamental is changed. The summary statistics in the new scale are easily computed from the original ones. We are simply choosing a different way to express the same thing. The second idea is that we can use an equation of the form $x^* = a + bx$ (or later $y = a + bx$) to express a relationship between two variables.

1.3 The Normal Distributions

Note that normal distributions are introduced here as a common model for the overall pattern of many sets of data, and not in the context of probability theory. Although this ordering of material is unusual, it has several advantages. The normal distributions appear naturally in the description of large amounts of data, so that the later assumption for inference that *the population has a normal distribution* becomes clearer. Moreover, mastering normal calculations at this point makes it easier to teach the material on probability (Chapters 5 and 6). If the students already know how to compute normal probabilities and have a fair understanding of the relative frequency interpretation from this section, the transition to general ideas about probability is facilitated. Of course, Chapter 6 presents additional facts about normal distributions in the context of probability.

The key idea is that we can use a mathematical model as an approximation to real phenomena. The 68-95-99.7 rule is a useful device for interpreting μ and σ for normal distributions. From the viewpoint of statistics, in contrast to that of probability, we always think of our models as approximations rather than the truth. This point can be illustrated by considering Example 1.22. The $N(430, 100)$ distribution is very useful for describing SAT scores of high school seniors. It can give reasonably accurate answers to interesting questions as illustrated in this example. However, It does not work very well in the extreme tails. For example, since the highest score possible is 800, the proportion of students scoring 830 or better is not appropriately calculated by using $Z \geq (830 - 430)/100 = 4$ and normal calculations. In a population of a million students, the calculation would give $1,000,000(.000317) = 317$ students scoring above 830!

Normal quantile plots are used frequently in the text. Students should learn to interpret them. If computer software is not available for constructing these plots, we do not recommend that they be drawn by hand. In this case, stemplots should be used to assess normality on a routine basis for the exercises.

Chapter 2 Looking at Data: Relationships

This chapter concludes the part of the text dealing with methods for looking at data. In the first chapter, techniques for studying a single variable were explored. Here the focus is on general methods for describing relationships between pairs of variables.

Scatterplots are used to examine pairs where both variables are quantitative. Details of least squares regression are presented. Correlation is introduced and its connection with regression is examined. Relations between categorical variables are described using proportions or percentages calculated from two-way tables. Issues related to the question of causation are illustrated by consideration of smoking and lung cancer.

The descriptive methods in this chapter, like those in earlier chapters, correspond to the formal inference procedures presented later in the text. By carefully describing data first, we avoid using inference procedures where they clearly do not apply. Fitting a least squares line is a general procedure, while using such a line to give a 95% prediction interval requires additional assumptions that are not always valid. In addition, students become accustomed to examining data *before* proceeding to formal inference, an important principle of good statistical practice. The data for a regression (Chapter 9) are displayed in a scatterplot. Side-by-side boxplots are useful with two-sample t-tests (Chapter 7) and analysis of variance (Chapter 10). The percentages obtained from two-way tables are the basis for the formal inferences on count data (Chapter 8).

2.1 Scatterplots

From the previous chapter the students will be familiar with the idea that graphical displays of data are useful. The extension to a scatterplot should reinforce this idea. The distinction between explanatory and response variables is a relatively simple distinction that is essential to the least squares method.

Constructing scatterplots is an easy task, particularly with a computer. Interpreting them, on the other hand, is an art that takes practice. For classroom discussion, you can use the examples given in the text or those presented in the exercises. Stress the idea that common sense and some understanding of the data are necessary to do a good job of description. Computers can make the plots, but people are needed to describe them. Again, the general rule is to look for overall patterns and deviations from them. Dichotomies such as positive and negative association are useful in many cases but can lead to distorted descriptions when imposed in situations where they do not apply.

The median trace is a useful data analytic tool for looking at some kinds of scatterplots. Try to avoid the type of thinking that says: if I have situation a, then use method b. A method is useful if it helps you to learn something about the data. Sometimes you do not know if it is useful until you try it.

Side-by-side boxplots can give a very informative data summary when one variable (usually the response variable) is quantitative and the other is categorical or concentrated on a small number of possible values. Since the students have already encountered boxplots, the idea of putting several of them alongside of each other should follow quite easily.

2.2 Least Squares Regression

The principle of least squares is introduced and the details of the computations are presented. As was mentioned in regard to the computation of the sample variance, hand calculations can be organized in a table. In this case the column headings are x, x^2, y, y^2, and xy.

Try not to burden the students with excessive computation. If a computer is available, the least squares line can be obtained from a regression routine. Note that the output may contain many pieces of information related to inference that are not relevant at this time. If computing facitlities are not available, assign exercises where most of the computations are given and the amount of arithmetic required is minimal.

Plotting the residuals versus x is very important. Again, construction of the plot is easy and can be done effectively by a computer, but interpreting the results is an art.

Outliers and influential observations can provide interesting class discussion. When a cause can be found or a lurking variable discovered, the value of carefully looking at data is reinforced. Formal rules for dealing with these situations are not advised. The point is that outliers and influential observations are present in many real data sets. A good analysis finds them and assesses their effects on the results.

2.3 An Application: Exponential Growth

The optional section on exponential growth introduces no new statistical concepts. The only difference from the previous section is that the function being fitted is more complex. Many students at this level have rather unpleasant thoughts when they hear the word logarithm. They are surprised to learn that logs can be viewed simply as that which is done to data to transform exponential growth into linear growth.

2.4 Correlation

The computations needed for a correlation are similar to those for a regression. Therefore, there is relatively little new here in this regard. A key point is that a correlation (or the square of a correlation) is a numerical summary of a relationship between two variables that are linearly related. Thus, as an aid to interpreting a scatterplot, it a potentially useful descriptor. With outliers, influential observations or nonlinear relationships. however, it may give a very distorted impression.

2.5 Relations in Categorical Data

The computation required in this section in minimal. Percents and proportions are the numerical summaries. On the other hand, some very important ideas are presented. Judgment is required to select what percents to calculate. The idea of conditional distributions appears here and can be emphasized if you intend to cover the optional material on conditional probability in Chapter 4. Simpson's paradox is not easily understood when first encountered. Careful class discussion of the examples in the text is needed if students are to grasp this idea.

2.6 The Question of Causation

The final section of this chapter is basically a case study concerning the relationship between smoking and lung cancer. There are no formulas but there are a lot of ideas. Class discussions may bring out some strong opinions of the part of the students. It is interesting to note that while there is a great deal of statistical information that is relevant to the issue, real conclusions regarding this very important question require judgement and expertise beyond the realm of formal statistical inference.

Chapter 3 Producing Data

This is a relatively short chapter with a lot of ideas and little numerical work. The message is that production of good data requires careful planning. Random digits (Table B) are used to assign units to treatments and to select simple random samples. There are several good examples that can serve as the basis for classroom discussion.

3.1 First Steps

This introductory section gives an overview of why careful planning of data collection is needed and makes the point that anecdotal evidence is not a good basis for drawing conclusions. Sampling and experimentation are described in general terms.

3.2 Design of Experiments

Terminology used in experimental design is introduced and is illustrated with examples. The advantages of comparative experiments are shown pointedly with the gastric freezing experiment (Example 3.5). Students seem to particularly enjoy this example and they often express outrage at the blatant misuse of the data described.

Once the need for comparison and randomization is established, methods for randomization of experimental units to treatments is presented. Note that there are many correct ways to do these randomizations. Many exercises using Table B are given. If you use statistical software, some particular randomization routines may be available. In very general terms, most randomizations can be accomplished by a random sort of the data. In any package that has a random number generator and a sort, this can be done by assigning a random uniform to each case and then sorting the resulting file. For example, to assign 10 people to each of two treatments, set up a file with cases corresponding to the integers 1 to 20. The assign a random number to each case and sort the file on the basis of the random numbers. The first 10 cases receive treatment one and the others receive treatment two. Note that this is not a particularly efficient algorithm but it is very clear and it helps the students understand the principle of randomization.

3.3 Sampling Design

The simple random sample (SRS) is described. Note that in later chapters the assumption that data or errors are normally distributed is expressed by saying that they are an SRS from a normal population.

Other sampling designs are mentioned, including stratified samples, multistage designs, and systematic random samples (Exercise 3.37). The problems of nonresponse and bias are also discussed.

3.4 Why Randomize

This section introduces some key ideas that will be further explained and used in later chapters. The idea of a parameter in a population contrasted with a statistic in a sample is fundamental. Through an opinion poll example, this distinction is made clear. The sampling distribution of the sample proportion is shown by simulation for samples of size 100 and 1785, thus illustrating the central limit theorem and the fact that the variability decreases as the sample size increases.

Because students often find probability hard and fail to see it's relevance, we have tried to motivate probability by first discussing randomization in data collection and its consequences. In teaching this section, try to get the students to think about and discuss the general ideas. This will facilitate their learning of the details to follow in later chapters.

Chapter 4 Probability: The Study of Randomness

In the last chapter it is explained that randomization is needed for the proper design of experiments and sampling plans. Therefore, the study of probability as a model for randomness naturally fits here.

An overview of the entire course is given in the following paradigm. In the first three chapters we studied how to look at and describe data. Thus, in a sense, we concentrated on the *sample*. Chapter 3 discussed how data are produced and introduced the idea of the *population* for which inference in desired. Now, we study the population in detail, with emphasis on the mechanisms by which samples are generated. In Chapters 6 to 10, we reverse the orientation and study the methods by which samples lead to inferences about populations.

This chapter and the one that follows will be difficult for many of the students. We have tried to emphasize a conceptual approach to the topics presented. Nonetheless, there are several rules and manipulations that must be mastered.

You should carefully consider how much depth and breadth you want to include for this material. By omitting the optional sections and treating some subjects (such as sample spaces) lightly, you can teach the minimum probability required for a proper understanding of statistical inference. On the other hand, the text provides sufficient material for more comprehensive coverage.

4.1 Probability Models

Sample spaces, assignment of probabilities and basic probability rules are introduced in this section. Try to emphasize that all of these ideas follow from common sense and are not an arbitrary mathematical system devoid of a intuitive foundation. Many of the students will have had unpleasant experiences with some of this material previously.

Many of the topics can be introduced through an example with the formalization coming later. Give a set of probabilities that have a sum greater than one and ask why this does not make sense. Then explain that their conclusion that the sum must be one is a basic rule of probability theory.

Similarly, most of the students can apply the addition rule and the complement rule to real problems without ever having encountered these rules formally. Let them figure out the rule first and then present it formally.

4.2 Random Variables

Random variables are the raw material for statistical inference. We therefore introduce them early in the presentation of probability. If an experiment or sample survey is repeated, different results will be obtained. The probability theory for random variables tells us that there is a certain type of regularity or predictability in these results.

4.3 Describing Probability Distributions

The mean, variance and standard deviation are familiar quantities from Chapter 1. Emphasize that they are a little different here because we are dealing with a population. If the students have a lot of difficulty with this transition, you can ask them to think about some very large samples. For example, in Example 4.18, think of an experiment with 10,000 trials in which the results are exactly as expected: 625 zeros, 2500 ones, 3750 twos, 2500 threes and 625 fours. The calculations can then be performed on this hypothetical large sample (in which the difference between n and $n-1$ is negligible) and the connection between probability and relative frequency can be reinforced.

In a similar way the rules for means and variances build on the parallel rules for sample statistics. This material can be treated briefly and reinforced with homework exercises.

The law of large numbers is the key concept in this section. It expresses a fundamental idea that is part of the foundation for much of statistical inference. Most of the students will have some intuitive idea about this law. Through class discussion these ideas can be explored and clarified.

4.4 Probability Laws

This optional section presents some of the traditional material on probability. The addition rules, multiplication rules and conditional probability are all discussed. The section is optional because these ideas are not essential to an understanding of the statistical ideas that follow in later chapters. Ample exercises are provided for the instructor who wants to cover this material.

Although the basic ideas contained in these probability laws are very intuitive, many students have a great deal of difficulty with the symbolic manipulations involved. By a detailed presentation of examples or exercises you can emphasize that these are calculations are based on common sense and are not simply a collection of abstract rules.

Chapter 5 From Probability to Inference

There are two key ideas in this chapter. First, a statistic is random variable, the value of which varies from experiment to experiment or sample to sample. Second, as a random variable, it has a sampling distribution with a mean and a standard deviation that can be calculated from the distribution of the basic random variables that are combined to calculate the statistic.

The binomial distribution is treated in detail. The normal approximation to the binomial is given as an important special case that sets the stage for the central limit theorem presented in the second section. The chapter concludes with an optional section on control charts where probability ideas are shown to be both useful and important.

5.1 Counts and Proportions

The amount of time you choose to spend on this section is quite flexible. The inference for count data presented in Chapter 8 is all based on the large sample normal approximation to the binomial. Therefore, the minimum coverage needed consists of the fact that the distribution of $\hat{p}$ is approximately normal with mean p and standard deviation $\sqrt{p(1-p)/n}$ when n is sufficiently large.

On the other hand, there is sufficient material for a full treatment of the binomial distribution. Table C gives individual probabilities for a variety of values of n and p and the formula for calculating binomial probabilities is explained in detail. Many computer routines are also available for calculating individual or cumulative binomial probabilities.

Try not to let the computational details interfere with the important pedagogical role played by this section. The students should understand that $\hat{p}$ is a random variable with a sampling distribution that is approximately normal for large n. In a given experiment or sample, only *one* realization or value of this statistic is obtained. Our inference will be based on this single observation and probability calculations will be based on an assumed (testing) or estimated (confidence intervals) sampling distribution for this statistic.

5.2 Sample Means

This is a rather short but important section where the mean and standard deviation for the sample mean are explained and the central limit theorem is presented. In terms of statistical applications, there are two key ideas. First, the central limit theorem is widely applicable and is the basis for normal based inference. Second, the standard deviation of the mean decreases as $\sqrt{n}$. This section completes the discussion of normal distributions begun in Chapter 1 by showing the normal distributions as sampling distributions.

5.3 Control Charts

This optional section illustrates how some probability calculations are used in practice. It is not intended as rigorous treatment of control charts. It can be used very effectively to help the students understand both probability and statistical inference. An out of control signal is unlikely to occur if the process is in control. Therefore, we *suspect* that the process has been disturbed if such a signal is observed.

Chapter 6 Introduction to Inference

This is a chapter with a lot of fundamental ideas. Confidence intervals and tests are presented with some cautions concerning the use and abuse of tests. Throughout, the setting is inference about the mean μ of a normal population with known standard deviation. As a consequence, the z procedures presented are not applicable to most real sets of data, but rather serve to ease the transition to the more useful procedures presented in the next chapter.

Experience has shown that many students will not master all of this material upon seeing it for the first time. Fortunately, they will see all of the key ideas again in the next chapter. By the time they have completed both chapters and worked many exercises, they should be able to grasp the fundamentals. Remember that mastery of the reasoning of inference is far more important that the number of procedures learned.

6.1 Estimating with Confidence

Confidence intervals are straightforward but easily misinterpreted. Try to emphasize the meaning rather than the details of computation. For most students this is accomplished more effectively by a discussion of Figure 6.2 or similar results than by an algebraic deviation of a formula for the coverage probability.

At this stage the students should be comfortable with the sample mean, population parameters versus sample statistics, the fact that $\bar{x}$ is approximately normal with mean μ and standard deviation $\sigma/\sqrt{n}$, and the table of normal probabilities (Table A). Stress the fact that we are simply putting these ideas together in this section to produce a useful statistical procedure. Similar comments hold for the next section.

The subsection on choosing the sample size reinforces the idea that the width of the interval depends upon n. This material is particularly important for students who will design studies in the future.

6.2 Tests of Significance

Although this section is not much longer than the previous one, it contains a few things that students generally find more difficult. The choice of one-sided versus two-sided tests and P-values versus fixed significance level testing are in this category. Try not to let these complications interfere with the presentation of the basic ideas.

Students seem to have the most difficulty when they become sufficiently frustrated that they put their common sense aside and start seeking formulas or rules that cover every possible case. Graphs illustrating the calculation of P-values are very useful here. Note that statistical software universally gives P-values. Therefore, an understanding of this approach to testing is essential for students who will either read or perform statistical studies.

6.3 Use and Abuse of Tests

The first two subsections of this section give some important ideas about significance testing that can serve as the basis for classroom discussion. Students are often reluctant to make judgements about the *importance* of an effect in contrast to its statistical significance. Emphasize that this judgement is an important part of using statistics with real data. Computers are easily programmed to perform calculations and produce P-values. Informed judgements are needed to translate and interpret these results.

The optional material on power is similar to the material on selecting the sample size for confidence intervals. It reinforces what has already been said about the dependence of the behavior of the procedure on the sample size and is very important for students who will be planning studies.

The last subsection on inference as a decision is also optional. It uses the traditional dichotomy between Type I and Type II errors and serves as a brief introduction to more general applications of decision theory.

Chapter 7 Inference for Distributions

The principles underlying confidence intervals and significance tests were presented in the previous chapter. We now apply these principles to inference problems for one and two samples. Throughout the chapter, we work under the realistic assumption that the population standard deviation is unknown.

The first section deals with inference for one sample. The problem of paired comparisons is treated as a special case. Optional sections on power and a nonparametric alternative to the normal based procedure are also given.

Two sample procedures are presented in the second section. The chapter concludes with an optional section on inference for the standard deviation.

From earlier chapters, the students should be familiar with looking at data from one or two samples carefully and with computing sample means and standard deviations. The additional arithmetic required for the construction of confidence intervals and significance tests is not particularly difficult. Many of the exercises provide means and standard deviations so that the students can concentrate on the new ideas presented here.

The transition from the normal to the t table requires some careful explanation. Because we are dealing with a family of distributions indexed by the degrees of freedom, only selected values can be given in a table. Since most of the exercises are based on real data, no attempt has been made to assure that an entry for the exact degrees of freedom for a particular problem are given in the table. Therefore, approximate values of t^* are to be used for confidence intervals. Similarly, bounds on the P-value can be obtained rather than exact P-values, using Table E.

A conservative approach for confidence intervals is to use a value for the degrees of freedom that is less than or equal to the required value. This results in intervals that are at least as wide as the exact interval, thereby ensuring that the coverage probability is at least as large as that specified. The same approach for tests ensures that the probability of a false rejection of H_0 is less than or equal to the value given. Table E provides sufficient detail so that for all practical purposes, inaccuracies resulting from these difficulties are very small. Students checking their solutions for odd-numbered exercises in the back of the text should be aware of these considerations. Of course, if a probability function for the t distribution is available in computer software, exact values are easily obtained. Note that it is common practice for many computer packages (and in reporting the results of a statistical analysis) to give P-values in the form $P <$ some number. Using Table E, when an extreme value of the t statistic is found, the P-value will be reported as $P < .0005$ for a one sided-test and $P < .001$ for a two-sided test.

7.1 Inference for the Mean of a Population

The term *standard error* is introduced for the estimated standard deviation of a statistic. Thus, t-statistics are normalized by dividing by a standard error, and confidence intervals are constructed by taking the value of a statistic plus or minus a constant times the standard error.

The usual normal theory one sample confidence intervals and significance tests are presented in this section. By first taking differences between pairs of observations, paired comparisons problems are treated in the same way. The idea of robustness is introduced and practical guidelines for using the t procedures are given.

The power of the t test is treated in an optional subsection. This material is very similar to that presented in the third section of Chapter 6 and does not make use of the non-central t distributions. Except for problems involving very small sample sizes, this method works quite well to produce sufficient accuracy to make judgements in practical situations. As was mentioned before, this material reinforces the fundamental idea that the performance characteristics of statistical procedures depend upon the sample size. It is important for students who will be planning statistical studies to be familiar with these ideas.

The final optional subsection discusses what can be done in some situations when the data are not normal. Transformations are mentioned and an explanation of the sign test for paired comparisons is given. The intention here is to point to the two major strategies available, and not to discuss them in detail.

7.2 Comparing Two Means

Procedures based on normal assumptions for comparing two means are presented in this chapter. To place the problems in a proper perspective, the situation where the standard deviations are known is discussed first.

When replacing the known standard deviations by sample estimates, difficulties regarding the appropriate degrees of freedom for the t distribution arise. Most computer software packages use an approximation similar to that given in Equation (7.4). This formula is given for information only. From an educational point of view, very little is accomplished by having students do a large amount of computation with this formula. In most practical situations use of the minimum of $n_1 - 1$ and $n_2 - 1$ as the degrees of freedom for the t gives essentially the same results. It is better to have the students concentrate on the appropriate use of the procedures and the interpretation of the results rather than to have them spend a lot of time with computation.

The pooled two-sample t procedures are presented in the optional last part of this section. Occasionally, students will question the usefulness of these procedures. Why assume that the standard deviations are equal when we have already learned procedures that are valid under less restrictive assumptions? From a theoretical point of view there are answers. We have exact rather than approximate distributions, confidence intervals will be shorter in some stochastic sense and significance tests have more power. On the other hand, from a practical point of view, there will be very little difference in the results (unless, of course, there are large differences in the standard deviations so that the pooled procedures are invalid). The primary reason for including this material is that it facilitates the introduction of similar ideas in Chapter 10, where analysis of variance is introduced. If you do not plan to cover analysis of variance and time is tight, you could easily omit this material.

7.3 Inference for Population Spread

The usual F test for comparing two variances is presented in this section. It is pointed out that this procedure is not robust with respect to the assumption of normality and a general discussion of robustness follows. This section can easily be omitted. The very poor performance of F in the two-sample variance setting should not be overlooked merely to teach a procedure for this case.

Note that this is the first place in the text that F distributions are used. They appear later in the chapters on regression (Chapter 9) and analysis of variance (Chapter 10). If you do not cover this section, you will need to spend a little more time explaining the tables when they are needed later in the text.

Chapter 8 Inference for Count Data

The first two sections of this chapter present the standard z procedures for one-sample and two-sample binomial problems. By now the students should be comfortable with the general framework for confidence intervals and significance tests. For those who have not yet completely mastered these concepts, this material affords an additional opportunity to learn these important ideas.

Inference for two-way tables is discussed in the last section. If desired, this material can be omitted without loss of continuity. The notation here is more complex than in earlier chapters. (There is no easy way to describe the general form of a two-dimensional array of data without two subscripts.)

8.1 Inference for a Single Proportion

Confidence intervals and significance tests for a single proportion are presented. A new complication that arises with these problems concerns the standard deviation of $\hat{p}$. For confidence intervals we use the standard error $\sqrt{\hat{p}(1-\hat{p})/n}$, whereas for tests we use $\sqrt{p_0(1-p_0)/n}$. Although this may seem confusing to the students at first, the basic idea is quite reasonable.

We use all of the information available in a problem for our calculations. For the confidence interval, p is assumed to be unknown and the standard error must therefore be estimated using the value of $\hat{p}$ obtained from the data. On the other hand, when testing $H_0 : p = p_0$, our calculations are

Chapter 8 Comments

based on the assumption that H_0 is true, and therefore, we use the value p_0 in the calculations. Note that these choices destroy the exact correspondence between confidence intervals and tests (reject if the hypothesized parameter value is outside of the confidence interval).

The section concludes with a description of the procedure for choosing a sample size to guarantee a given bound on the width of a confidence interval. From a practical point of view, the table given in Example 8.7 is very informative. The widths of intervals vary relatively little for values of $\hat{p}$ between .3 and .7.

8.2 Comparing Two Proportions

Confidence intervals and significance tests for comparing two proportions are presented in this section. From Chapter 7, students should be familiar with the basics of two-sample problems. As in the previous section, we use different standard errors for confidence intervals and tests. The idea of pooling information from two samples to estimate a standard error was first presented in the optional part of the second section of Chapter 7. This principle will be used again in Chapter 10 when we treat analysis of variance.

8.3 Inference for Two-Way Tables

You have some choice regarding depth of coverage for this section. As was mentioned above, this section can be eliminated entirely if desired. The computations for the X^2 statistic are rather straightforward although a fair amount of arithmetic is required and care must be taken to ensure accurate results. The vague idea of testing for a relationship between two categorical variables is also fairly straightforward. Therefore, it is possible to teach this section spending a minimum amount of time on the two models described.

On the other hand, discussion of the models in detail gives the students a good background for learning about model based inference for more complex situations. The statistical model gives a clear statement of the assumptions needed for a given procedure and specifies the parameters about which inference is desired. A major part of statistical inference concerns the translation of a vague notion, such as dependence between two categorical variables, into a testable hypothesis stated in terms of the parameters of a statistical model. To draw meaningful conclusions, the results of the analysis, stated in terms of the model, must then be translated back into the context of the real problem.

Chapter 9 Inference for Regression

The two sections of this chapter present simple linear regression and multiple regression. Having completed Chapter 2, the students should be familiar with the data analytic issues that arise with simple linear regression. Therefore, the first section is primarily concerned with applying statistical inference principles to this problem. The material from Chapter 7 is particularly important in this regard. On the other hand, nothing in Chapter 8 is needed for an understanding of this chapter.

The material on multiple regression is presented in general terms with little detail regarding computation. A detailed presentation of the case study in class will give the students a good overview of many of the important considerations needed for the proper use of this statistical method.

9.1 Simple Linear Regression

Note that there are many interesting problems for which the relationship between two variables can be summarized graphically and numerically with a least squares line. Not all of these can be analyzed using the methods presented in this chapter. Inference for linear regression is based on a statistical model that expresses the assumptions underlying the inference procedures.

Confidence intervals and significance tests should be familiar concepts for the students at this point. Although the computational details are more difficult in this section, the underlying principles are not new. Use of statistical software is very effective here. It allows the students to concentrate on learning the concepts without being unduly concerned with a long sequence of computations leading to a numerically correct answer. Stress that the number is not the *answer*; the interpretation of the number in the context of a real problem is the proper end result of a statistical analysis.

Prediction intervals are a new idea. By contrasting them with confidence intervals, the students can avoid common misinterpretations of confidence intervals.

Most computer packages provide an analysis of variance table as part of the output for a regression. For simple linear regression the table is not particularly useful. Often it generates more confusion than useful information for students at this point. On the other hand, the anova table is very important for multiple regression and it is essential for performing analysis of variance. If you plan to cover these topics, discussion of the anova table for linear regression provides an excellent introduction to this topic. Explain that it is a different way to present some of the information already discussed and that it will be important for topics to be covered later.

The section concludes with optional material on inference for correlation. It is pointed out that the test for a zero correlation is equivalent to the test for a zero slope in the regression. The square of the correlation is expressed in a form that generalizes easily to the multiple regression setting.

9.2 Multiple Regression

The model for multiple regression and a brief overview of the inference procedures are presented. This section can be omitted without loss of continuity. For students who will take another statistics course in which this material will be covered in great detail, a discussion of the case study can be very valuable. Many important ideas are presented in the context of the problem of predicting grade point average from high school grades and SAT scores.

Chapter 10 Analysis of Variance

The two sections in this chapter present one-way and two-way analysis of variance. The one-way problem is seen as a generalization of the two-sample case presented in Chapter 7. With the generalization come several new ideas, however.

Two-way analysis of variance is treated conceptually. A major goal is for the students to understand the meaning of interaction. No computational formulas are given. The emphasis is on understanding the assumptions and interpreting the output.

10.1 One-Way Analysis of Variance

The idea of using a statistical test to compare population means is familiar from Chapter 7. New issues arise because we may have more than two means to compare. The DATA = FIT + RESIDUAL idea is used in presenting the statistical model underlying the analysis. The model is treated first, thereby emphasizing the principle that inference is based upon assumptions and definitions of parameters given by a model.

The analysis of variance table is constructed using some familiar ideas. Pooling of variances was treated in Chapter 8 and again the idea of DATA = FIT + RESIDUAL is used. The anova tables from the regression chapter (Chapter 9) anticipated many of the ideas presented here.

Some new ideas arise with contrasts and multiple comparisons. Without some discussion of at least one of these topics, one is left with a statistical procedure giving very little in terms of useful interpretable results. Most statistical software packages will perform these computations.

The issue of what multiple comparison procedure to use is rather difficult. We have chosen to present only one procedure—the Bonferroni. However, it is presented in such a way that any other procedure could easily be substituted.

The optional section on power reminds us again that the performance of a statistical procedure depends upon the sample sizes and the true values of the parameters. It is interesting to note that these kinds of calculations are rather easy to perform with any statistical software that has a function for the noncentral F distribution. Tables of power and even graphs of power functions are relatively easy to generate.

10.2 Two-Way Analysis of Variance

This section is really an introduction to more complex analysis of variance designs. Emphasis is on general ideas with computation left for the computer. In this context, the initial discussion of advantages of two-way anova is particularly important. Again the description of the model plays a key role.

Interaction is the important new idea. It is illustrated with a variety of examples chosen to demonstrate the different types of interpretations that arise in practice. Stress that the statement that there is interaction between two factors is essentially meaningless in itself. Interpretation requires a careful examination of the means in the context of a given problem.

3 SAMPLE EXAMS

3.1 SAMPLE EXAMINATION 1

This sample examination covers the material of Chapters 1 to 2 and illustrates the types of questions that can be asked. It is too long for a normal class period.

PART I–MULTIPLE CHOICE–3 POINTS EACH

1. The heights of American men aged 18 to 24 are approximately normally distributed with mean 68 inches and standard deviation 2.5 inches. Half of all young men are shorter than
(a) 65.5 inches
(b) 68 inches
(c) 70.5 inches
(d) can't tell, because the median height is not given

2. Use the information in the previous problem. Only about 5% of young men have heights outside the range
(a) 65.5 inches to 70.5 inches
(b) 63 inches to 73 inches
(c) 60.5 inches to 75.5 inches
(d) 58 inches to 78 inches

3. Use the information in Problem 1. What percent of young men are taller than 6 feet?
(a) 94.5%
(b) 44.5%
(c) 5.5%
(d) 2.5%

4. A study found correlation $r = .61$ between the sex of a worker and his or her income. You conclude that
(a) women earn more than men on the average
(b) women earn less than men on the average
(c) an arithmetic mistake was made; this is not a possible value of r
(d) this is nonsense because r makes no sense here

5. The scores on a statistics exam are strongly skewed to the left. So it is best to describe the distribution by reporting
(a) the five-number summary
(b) the mean and standard deviation
(c) the mean, median, and mode
(d) the correlation and its square

6. The grade point averages (GPA) of 7 randomly chosen students from your statistics class are

$$3.14 \ 2.37 \ 2.94 \ 3.60 \ 1.70 \ 4.00 \ 1.85$$

The mean GPA for these students is
(a) 2.8
(b) 2.94
(c) 3.6
(d) none of the above

7. Refer to the information given in the previous problem. If $\sum(x - \bar{x})^2 = 4.51$, then the standard deviation is
(a) .75
(b) .87
(c) .64
(d) .80
(e) none of the above

8. You record the age, marital status, and earned income of a sample of 1463 women. The number of variables you have recorded is
(a) 1463
(b) four—age, marital status, income, and number of women
(c) three—age, marital status, and income
(d) two—age and income. Marital status is not a variable because it doesn't have a unit like dollars or years.

9. A copy machine dealer has data on the number x of copy machines at each of 89 customer locations and the number y of service calls in a month at each location. Summary calculations give $\bar{x} = 8.4$, $s_x = 2.1$, $\bar{y} = 14.2$, $s_y = 3.8$, and $r = .86$. What is the slope of the least squares regression line of number of service calls on number of copiers?

3.1 SAMPLE EXAMINATION 1

(a) 0.86
(b) 1.56
(c) 0.48
(d) none of these
(e) can't tell from the information given

10. In the setting of the previous problem, about what percent of the variation in number of service calls is explained by the linear relation between number of service calls and number of machines?
(a) 86%
(b) 93%
(c) 74%
(d) none of these
(e) can't tell from the information given

11. A machine is designed to fill 16 ounce bottles of shampoo. When the machine is in control, the the mean amount poured into the bottles is 16.05 ounces with a standard deviation of .05 ounces. To make a control chart for this machine you would put the control limits at
(a) 16.05 and 32.10 ounces
(b) -.10 and +.10 ounces
(c) 16.00 and 16.10 ounces
(d) 15.95 and 16.15 ounces
(e) none of the above

12. A standardized test designed to measure math anxiety has a mean of 100 and a standard deviation of 10 in the population of first year college students. Which of the following observations would you suspect is an outlier?
(a) 150
(b) 100
(c) 90
(d) all of the above
(e) none of the above

13. Erin is a runner who keeps accurate records of her training. In one season she averaged 3.4 miles per day with a standard deviation of .5 miles. What is her standard deviation expressed in kilometers? (1 mile = 1.6 kilometers)
(a) .80 kilometers

(b) 2.13 kilometers
(c) 5.44 kilometers
(d) .31 kilometers
(e) none of the above

14. In a statistics course a linear regression equation was computed to predict the final exam score from the score on the first test. The equation was $\hat{y} = 10 + .9x$ where y is the final exam score and x is the score on the first exam. Eve scored 95 on the first test. What is the predicted value of her score on the final exam?
(a) 95
(b) 85.5
(c) 90
(d) 95.5
(e) none of the above

15. Refer to the previous problem. On the final exam Eve scored 98. What is the value of her residual.
(a) 98
(b) 2.5
(c) -2.5
(d) 0
(e) none of the above

PART II
PROBLEMS 1 AND 3—20 POINTS EACH;
PROBLEM 2—15 POINTS

1. The table below gives the percent of public school students in each state who graduate from high school. (Department of Education data for 1986, reported in *USA Today*, March 2, 1988.)

State	%	State	%	State	%	State	%	State	%
Ala	67.3	Ga	62.7	Md	76.6	N.J.	77.6	S.C.	64.5
Alaska	68.3	Hawaii	70.8	Mass	76.7	N.M.	72.3	S.D.	81.5
Ariz	63.0	Idaho	79.0	Mich	67.8	N.Y.	64.2	Tenn	67.4
Ark	78.0	Ill	75.8	Minn	91.4	N.C.	70.0	Texas	64.3
Calif	66.7	Ind	71.7	Miss	63.3	N.D.	89.7	Utah	80.3
Colo	73.1	Iowa	87.5	Mo	75.6	Ohio	80.4	Vt	77.6
Conn	89.8	Kan	81.5	Mont	87.2	Okla	71.6	Va	73.9
Del	70.7	Ky	68.6	Neb	88.1	Ore	74.1	Wash	75.2
D.C.	56.8	La	62.7	Nev	65.2	Penn	78.5	W.V.	75.2
Fla	62.0	Me	76.5	N.H.	73.3	R.I.	67.3	Wis	86.3
								Wyo	81.2

(a) Make a stemplot of the distribution of graduation rates. Describe the overall shape of the distribution. Are there any outliers? If there are outliers, suggest an explanation.

(b) Give the five-number summary for this distribution.

(c) Use the $1.5 \times IQR$ criterion to spot suspected outliers. Does this criterion point out any outliers that you identified by eye in (a)?

2. In a study of the change in social change, you obtain data from the National Center for Health Statistics for the life expectancy at birth of people born in the United States between 1920 and 1985. Life expectancy increased from 54.1 years to 74.7 years during this period. The growth in life expectancy was roughly linear ($r^2 = .94$), so you fit the least squares line to the growth data. Here is the *residual plot* from this least squares fit.

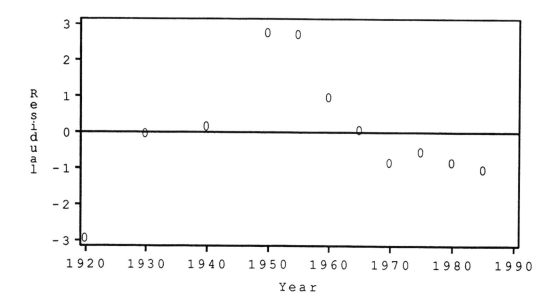

Describe carefully what the pattern of the residuals tells you about the growth of life expectancy over time.

3. The presence of sulfur compounds in wine is a defect that causes wine lovers to describe the odor of the wine as "cabbage" or "onion." Wine producers are therefore interested in the "odor threshold," the lowest concentration of these compounds that can be detected by trained judges. In a threshold determination experiment, a judge is presented with three glasses. One is identified as containing the wine alone for reference. The remaining two glasses are not identified: one contains the wine, and in the other the wine has been spiked with a known concentration of the sulfur compound. The judge tries to identify the spiked glass. Each judge is presented with five concentrations of the sulfur compound in increasing order.

Twenty judges took part in one such trial. Here are the results for the compound ethanethiol. The explanatory variable is the logarithm of the ethanethiol concentration (in micrograms per liter) in the spiked wine. The response variable is the percent of the 20 judges who correctly identified the spiked wine. (Simplified from O. J. Goniak and A. C. Noble, "Sensory study of selected volatile sulfur compounds in white wine," *American Journal of Enology and Viticulture*, 38(1987), pp. 223–227.)

3.1 SAMPLE EXAMINATION 1

Log concentration	0.3	0.5	0.6	0.7	0.8
% correct	55	65	75	70	85

(a) Experience has shown that the percent of correct responses usually increases linearly with the logarithm of the concentration. Plot the data, evaluate the linearity of the relationship, and look for outliers and influential observations.

(b) Compute the least-squares regression line and draw it on your plot.

(c) The odor threshold is determined by using this regression line to find the concentration that is identified correctly 75% of the time. What is the odor threshold for ethanethiol? (Notice that this is not the usual task of predicting y for a specified x. There are more elaborate statistical methods for finding the x that corresponds to a specified y in regression, but these methods were not used by the wine experts.)

ANSWERS TO SAMPLE EXAMINATION 1

PART I

1. b. 2. b. 3. c. 4. d. 5. a. 6. a. 7. b. 8. c. 9. b. 10. c. 11. d. 12. a. 13. a. 14. d. 15. b.

PART II

1. (a)

```
5 | 6
6 | 22233444
6 | 56777788
7 | 0001123334
7 | 555566677889
8 | 00111
8 | 677899
9 | 1
```

(b) The five number summary is 56, 67, 73, 79, 91.

(c) The inter quartile range is 79-67=12. 1.5 times IQR is 18. The median plus or minus 1.5 times IQR is 55 to 91. There are no outliers. Note that the observation 91 is exactly 1.5IQR above the median.

2. The residuals show a clear pattern of negative values, then positive values, then negative values. Relative to the linear pattern, life expectancy was less than predicted in 1920, more than expected from 1940 to 1960, and a little less than expected after 1960.

3.1 SAMPLE EXAMINATION 1

3. (a) The relationship is approximately linear. There are no clear outliers or influential observations. (b) $\hat{y} = 38.6 + 54.1x$ (c) $x = .67$

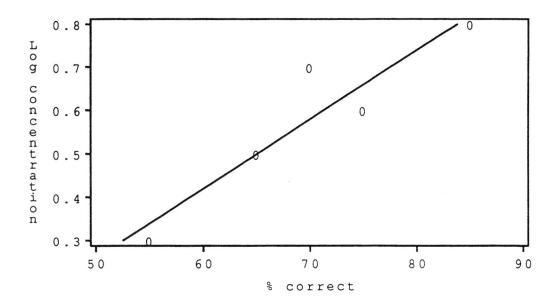

3.2 SAMPLE EXAMINATION 2

This sample examination covers the material of Chapters 3 to 5 and illustrates the types of questions that can be asked. It is too long for a normal class period.

PART I–MULTIPLE CHOICE–3 POINTS EACH

1. You notice that your car seems to run better when you use Brand A of gasoline than when you use Brand B. Can you conclude that Brand A is better than Brand B for your car?
(a) Yes. This is a simple random sample
(b) No. The evidence is anecdotal
(c) Yes. This is a comparative experiment
(d) No. The two brands are the same
(e) None of the above

2. A new headache remedy is given to a group of 25 patients who suffer severe headaches. Of these, 20 report that the remedy is very helpful in treating their headaches. From this information you conclude
(a) the remedy is effective for the treatment of headaches.
(b) nothing, because the sample size is too small
(c) nothing, because there is no control group for comparison
(d) the new treatment is better than aspirin
(e) none of the above

3. A student organization wants to assess the attitudes of students toward a proposed change in the hours that the library is open. They randomly select 50 freshmen, 50 sophomores, 50 juniors and 50 seniors. The situation described is
(a) a stratified random sample
(b) a simple random sample
(c) a comparative experiment
(d) anecdotal evidence
(e) none of the above

4. In a study of the effects of acid rain, a random sample of 100 trees from a particular forest are examined. Forty percent of these show some signs of

3.2 SAMPLE EXAMINATION 2

damage. Which of the following statements are correct?
(a) 40% is a parameter
(b) 40% is a statistic
(c) 40% of all trees in this forest show signs of damage
(d) more than 40% of the trees in this forest show some signs of damage
(e) less than 40% of the trees in this forest show some signs of damage

5. Refer to the previous exercise. Which of the following statements are correct?
(a) The sampling distribution of the proportion of damaged trees is approximately normal
(b) If we took another random sample of trees, we would find that 40% of these would show some signs of damage
(c) If a sample of 1000 trees was examined the variability of the sample proportion would be larger than for a sample of 100 trees
(d) This is a comparative experiment
(e) none of the above

6. A randomly selected student is asked to respond yes, no or maybe to the question: "Do you intend to vote in the next presidential election?" The sample space is { yes, no, maybe }. Which of the following represent a legitimate assignment of probabilities for this sample space?
(a) .4, .4, .2
(b) .4, .6, .4
(c) .3, .3, .3
(d) .5, .3, −.2
(e) none of the above

7. In a population of students, the number of calculators owned is a random variable X with $P(X = 0) = .2$, $P(X = 1) = .6$, and $P(X = 2) = .2$. The mean of this probability distribution is
(a) 0
(b) 2
(c) 1
(d) .5
(e) the answer cannot be computed from the information given

8. Refer to the previous problem. The variance for this probability distribution is
(a) 1
(b) .63
(c) .5
(d) .4
(e) the answer cannot be computed from the information given

9. You play tennis regularly with a friend and from past experience you believe that the outcome of each match is independent. For any given match you have a probability of .6 of winning. The probability that you win the next two matches is
(a) .36
(b) .6
(c) .4
(d) .16
(e) 1.2

10. The number of calories is a one ounce serving of a certain breakfast cereal is a random variable with mean 110. The number of calories in a full cup of whole milk is a random variable with mean 140. For breakfast you eat one ounce of the cereal with 1/2 cup of whole milk. Let Z be the random variable that represents the total number of calories in this breakfast. The mean of Z is
(a) 110
(b) 140
(c) 180
(d) 250
(e) 195

11. In a large population of college students, 20% of the students have experienced feelings of math anxiety. If you take a random sample of 10 students from this population, the probability that exactly 2 students have experienced math anxiety is
(a) .3020
(b) .2634
(c) .2013

(d) .5
(e) 1

12. Refer to the previous problem. The standard deviation of the sample proportion of students who have experienced math anxiety is
(a) .0160
(b) .1265
(c) .2530
(d) 1
(e) .2070

13. If a population has a standard deviation σ, then the standard deviation of mean of 100 randomly selected items from this population is
(a) σ
(b) 100σ
(c) $\sigma/10$
(d) $\sigma/100$
(e) .1

14. The central limit theorem states that
(a) the sample mean is unbiased
(b) the sample mean is approximately normal
(c) the binomial distribution is skewed
(d) the sample standard deviation is approximately normal
(e) none of the above

15. An opinion poll asks a random sample of voters, "Do you think elected government officials are underpaid?" Suppose 25% of the population would respond "yes." If the sample size is 400, the probability that at least 90 respond "yes" is approximately
(a) .875
(b) .125
(c) .750
(d) .225
(e) none of the above

PART II
PROBLEM 1—15 POINTS;
PROBLEMS 2 AND 3—20 POINTS EACH

1. An advertisement claims that by taking a particular training course, significant increases in SAT scores result. Outline the design of an experiment that you would use to examine this claim. Assume that you have 200 students to use in the experiment.

2. Patients receiving artificial knees often experience pain after surgery. The pain is measured on a subjective scale with possible values or 1 to 5. Assume that X is a random variable representing the pain score for a randomly selected patient. The following table gives part of the probability distribution for X.

x_i	1	2	3	4	5
p_i	.1	.2	.3	.3	

(a) Find $P(X = 5)$.
(b) Find the probability that the pain score is less than 3.
(c) Find the mean μ for this distribution.
(d) Find the variance σ^2 for this distribution.
(e) Find the standard deviation σ for this distribution.
(f) Suppose the pain scores for two randomly selected patients are recorded. Let Y be the random variable representing the sum of the two scores. Find the mean of Y.
(g) Find the standard deviation of Y.

3. Assume that 10% of a population of electronic components are defective. For this problem, consider drawing a random sample of n components from this population.
(a) Give the probability distribution for the number of defectives when $n = 2$.
(b) Give the mean and standard deviation for the number of defectives when $n = 500$.
(c) Give the mean and standard deviation for the proportion of defectives when $n = 500$.
(d) What is an approximate value for the probability that the proportion of

defectives is greater than .1 when $n = 500$?

(e) What in approximate value for the probability that the number of defectives is less than 40 when $n = 500$?

ANSWERS TO SAMPLE EXAMINATION 2

PART I

1. b. 2. c. 3. a. 4. b. 5. a. 6. a. 7. c. 8. d. 9. a. 10. c. 11. a. 12. b. 13. c. 14. b. 15. a.

PART II

1. A control group of students who do not take the training course is needed. Randomly assign the 200 students into two groups of 100 each. One group takes the course while the other does not. All students should take the SAT at the same time under the same conditions. The statement about increases in the SAT scores could be interpreted in different ways. One interpretation would suggest that the SAT should be administered before and after the time that the training course takes place. With this interpretation, the changes in scores would be compared.

2. (a) $P(X = 5) = .1$. (b) $P(X < 3) = .3$. (c) $\mu = 3.1$. (d) $\sigma^2 = 1.29$. (e) $\sigma = 1.14$. (f) $\mu(Y) = 6.2$. (g) $\sigma(Y) = 1.61$.

3. (a) $P(X = 0) = .81$, $P(X = 1) = .18$, $P(X = 2) = .01$. (b) $\mu = 50$, $\sigma = 6.7$. (d) .5. (e) .0681.

3.3 SAMPLE FINAL EXAM

The following is a sample comprehensive final exam. It emphasizes the material from Chapters 6 to 8 and the first section of Chapter 9.

PART I–MULTIPLE CHOICE–4 POINTS EACH

1. You want to compute a 95% confidence interval for a population mean. Assume that the population standard deviation is known to be 10 and the sample size is 50. The value of z^* to be used in this calculation is
(a) 1.645
(b) 2.009
(c) 1.960
(d) .8289
(e) .8352

2. You want to estimate the mean SAT score for a population of students with a 90% confidence interval. Assume that the population standard deviation is $\sigma = 100$. If you want the margin of error of the to be approximately 5, you will need a sample size of
(a) 16
(b) 271
(c) 38
(d) 1476
(e) none of the above

3. You have measured the systolic blood pressure of a random sample of 25 employees of a company located near you. A 95% confidence interval for the mean systolic blood pressure for the employees of this company is (122, 138). Which of the following statements gives a valid interpretation of this interval?
(a) 95% of the sample of employees have a systolic blood pressure between 122 and 138.
(b) 95% of the population of employees have a systolic blood pressure between 122 and 138.
(c) If the procedure were repeated many times, 95% of the resulting confidence intervals would contain the population mean systolic blood pressure.

(d) The probability that the population mean blood pressure is between 122 and 138 is .95.
(e) If the procedure were repeated many times, 95% of the sample means would be between 122 and 138.

4. A significance test was performed to test the null hypothesis $H_0 : \mu = 2$ versus the alternative $H_a : \mu \neq 2$. The test statistic is $z = 1.40$. The P-value for this test is approximately
(a) .16
(b) .08
(c) .003
(d) .92
(e) .70

5. A significance test gives a P-value of .04. From this we can
(a) reject H_0 with $\alpha = .01$
(b) reject H_0 with $\alpha = .05$
(c) say that the probability that H_0 is false is .04.
(d) say that the probability that H_0 is true is .04.
(e) none of the above

6. You want to compute a 90% confidence interval for the mean of a population with unknown population standard deviation. The sample size is 30. The value of t^* you would use for this interval is
(a) 1.96
(b) 1.645
(c) 1.699
(d) .90
(e) 1.311

7. A 95% confidence interval for the mean reading achievement score for a population of third grade students is (44.2, 54.2).
The margin of error of this interval is
(a) 95%
(b) 5
(c) 2.5
(d) 54.2
(e) the answer cannot be determined from the information given

3.3   SAMPLE FINAL EXAM   59

8. Refer to the previous problem. The sample mean is
(a) 44.2
(b) 54.2
(c) .95
(d) 49.2
(e) the answer cannot be determined from the information given

9. Using the same set of data, you compute a 95% confidence interval and a 99% confidence interval. Which of the following statements is correct?
(a) the intervals have the same width
(b) the 99% interval is wider
(c) the 95% interval is wider
(d) you cannot determine which interval is wider unless you know n and s

10. To use the two-sample t procedure to perform a significance test on the difference between two means, we assume
(a) the population standard deviations are known
(b) the samples from each population are independent
(c) the distributions are exactly normal in each population
(d) the sample sizes are large
(e) all of the above

11. In an opinion poll, 25% of 200 people sampled said that they were strongly opposed to having a state lottery. The standard error of the sample proportion is approximately
(a) .03
(b) .25
(c) .00094
(d) 6.12
(e) .06

12. You want to design a study to estimate the proportion of students on your campus who agree with the statement, "The student government is an effective organization for expressing the needs of students to the administration." You will use a 95% confidence interval and you would like the margin of error of the interval to be .05 or less. The minimum sample size requires is approximately
(a) 22

(b) 1795
(c) 385
(d) 271
(e) none of the above

13. A two-way table of counts is analyzed to examine the hypothesis that the row and column classifications are independent. There are 3 rows and 4 columns. The degrees of freedom for the X^2 statistic are
(a) 12
(b) 6
(c) $n-1$
(d) the minimum of $n_1 - 1$ and $n_2 - 1$
(e) none of the above

14. A simple linear regression was used to predict the score y on a final exam from the score x on the first exam. The slope of the least squares regression line is .75. The standard error of the slope is .11 and the sample size is 200. A 90% confidence interval for the true slope is
(a) .64 to .86
(b) .53 to .97
(c) .57 to .93
(d) −.64 to .86
(e) none of the above

15. Refer to the previous problem. To test the null hypothesis that the slope is zero versus the one-sided alternative that the slope is positive, we use the statistic $t =$
(a) 6.82
(b) .05
(c) .15
(d) .95
(e) .75

PART II
PROBLEMS 1 AND 2—15 POINTS EACH;
PROBLEM 3—10 POINTS

1. In an experiment designed to compare the effectiveness of two methods of teaching French, 20 students were randomly assigned to each of the methods. The scores on a final exam are to be compared. A summary of the results is given in the following table.

	n	$\bar{x}$	s
Method A	20	82	12
Method B	20	77	14

For this problem, do not assume that the two population standard deviations are the same.
(a) State appropriate null and alternative hypotheses for comparing the two methods.
(b) Calculate the test statistic for the comparison.
(c) What is the approximate distribution of the test statistic under the assumption that the null hypothesis is true?
(d) Give an approximate P-value for the significance test.
(e) What do you conclude?

2. A study was conducted to examine the reasons why consumers purchase items from catalogs. From a computer list of people who were known to have made purchases from catalogs during the past year, questionnaires were sent to a sample. Two hundred questionnaires were returned. One question asked whether or not price was an important factor in the decision to make a purchase from a catalog. Another question asked the same about convenience. The counts for these two questions are given in the following table.

		PRICE	
		Yes	No
CONVENIENCE	Yes	100	50
	No	20	30

A question of interest is whether or not the responses to these two questions are independent.
(a) Calculate a table of expected counts under the assumption that the responses to the two questions are independent.
(b) What is the value of the statistic used for testing independence in this problem?
(c) What is the approximate distribution of this statistic under the null hypothesis of independence?
(d) Find an approximate P-value for the significance test.
(e) What do you conclude?

3. A study is performed to investigate the relationship between college and high school grade point averages. A simple linear regression is used to predict the first year grade point average of college students using their high school grade point average. A sample of 100 students is used for the study. The equation of the least squares line fit to the data is $y = 1.24 + .65x$.
(a) If the standard error of the slope is .04, find a 95% confidence interval for the slope of the true line.
(b) Test the null hypothesis that the slope of the line is zero versus that alternative that it is positive. Give the test statistic, an approximate P-value and state your conclusion.

ANSWERS TO SAMPLE FINAL EXAM

PART I

1. c. 2. b. 3. c. 4. a. 5. b. 6. c. 7. b. 8. d. 9. c. 10. b. 11. a. 12. c. 13. b. 14. c. 15. a.

PART II

1. (a) $H_0 : \mu_1 = \mu_2$; $H_a : \mu_1 \neq \mu_2$. (b) $t = 1.21$. (c) $t(19)$, the t distribution with 19 degrees of freedom. (d) $.2 < P < .3$. (e) There is no clear evidence to conclude that the two methods of teaching give different mean final exam scores.

2. The expected counts are 90, 60, 30 and 20. (b) $X^2 = 11.11$. (c) $\chi^2(1)$. (d) $P < .001$. (e) Responses to the two questions are dependent. Consumers who think price is important also tend to think that convenience is important.

3. .57 to .73. (b) $t = 16.25$, $df = 98$, $P < .001$. There is evidence to conclude that the slope is positive. There is a positive relationship between the high school grade point average and the first year college grade point average.

SOLUTIONS TO EXERCISES

3.4 CHAPTER 1

1.1 There are many correct answers to this problem. Size dimensions such as height, width, and depth in inches measured with a ruler. Volume in cubic inches. Weight in ounces measured with a scale. Number of pages, number of lines, and number of words could be counted. Number of words would relate to reading time while the size variables would tell if it fits into a book bag.

1.2 A tape measure can be used to measure in inches various lengths such as longest single hair, length of hair on sides or back or front. Details of how to measure should be given. The case of a bald person would make an interesting class discussion.

1.3 Population, population density, number of automobiles, various measures of air quality, commuting times, parking availability, cultural activities, cost of homes, taxes, quality of schools. Reasons should be given.

1.4 If the number of people with the disease is increasing, the number of deaths could increase even though treatment is improving. A rate is appropriate here.

1.5 Weight because it is more sensitive.

1.6 Rates are appropriate. A: .21, B: .40.

1.7 The one minute period should give the best results because it is based on more information. Some discussion is possible here. If the pulse is changing over time, for example, after exercise, the one minute period might not be the best.

1.8 It is difficult to count fractions of a beat so the measure is truncated. The problem is less with the longer intervals. If the time in seconds for 80 beats is t, the conversion is $60 \times (80/t)$. Because the truncation problem is not present with this method it is better.

1.9 Possible answers are total profits, number of employees, total value of stock, and total assets.

1.10 Total weight of the roots, height, girth or circumference 4 feet from the ground.

1.11 1120, 1001, 1017, 982, 989, 961, 960, 1089, 987, 976, 902, 980, 1098, 1057, 913, 999

1.12 Move the decimal six places to the right. 1084, 1131, 887, 639, 1216, 903, 977, 1088, 940, 1069, 667, 536.

1.13 Figure 1.6(a) is strongly skewed to the right with a peak at 0; Figure 1.6(b) is somewhat symmetric with a central peak at 4. The peak is at the lowest value in Figure 1.6(a) and is at a central value in Figure 1.6(b).

1.14 The distribution is skewed to the right with a single peak. There are no gaps or outliers.

1.15 There are two peaks. The ACT states are located in the upper portion of the distribution.

1.16 The distribution is roughly symmetric There are peaks at 240, 290, and 300. The middle of the distribution is around 270.

1.17 200 appears to be an outlier. The median is approximately 140.

```
10 | 139
11 | 5
12 | 669
13 | 77
14 | 08
15 | 244
16 | 55
17 | 8
18 |
```

19 |
20 | 0

1.18 The preference is subjective. With the split stems there is a suggestion of a possible second peak. It is roughly symmetric with a center around 11.0. Alaska (AK) and Florida (FL) appear to be outliers.

```
                                    3 |
                                    3 | 8
                                    4 |
                                    4 |
                                    5 |
                                    5 |
                                    6 |
                                    6 |
                                    7 |
                                    7 |
 3 | 8                              8 | 4
 4 |                                8 |
 5 |                                9 |
 6 |                                9 | 459
 7 |                               10 | 034
 8 | 4                             10 | 667899
 9 | 459                           11 | 3
10 | 034667899                     11 | 677889
11 | 3677889                       12 | 2234
12 | 2234555688                    12 | 555688
13 | 001244557888                  13 | 001244
14 | 036799                        13 | 557888
15 |                               14 | 03
16 |                               14 | 6799
17 | 8                             15 |
                                   15 |
                                   16 |
                                   16 |
                                   17 |
                                   17 | 8
```

1.19 (a) The distribution is somewhat skewed to the right. There is one peak. The center is 523 eggs. 915 and 945 appear to be unusual values.

```
3 | 99
4 | 01123345789
5 | 24457889
6 | 5
7 | 058
8 |
9 | 14
```

(b) The egg production increases up to day 7; then falls off to day 12. From day 13 to day 27, there is a small decreasing trend.

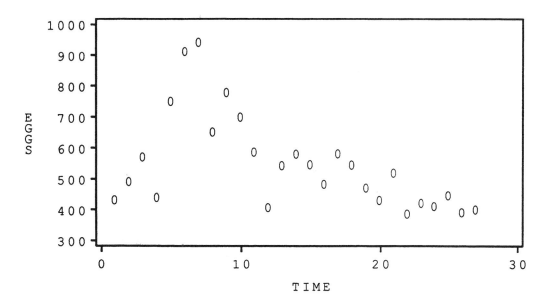

1.20 (a) The stems should be split. The distribution is roughly symmetric with one high outlier.

```
1 | 124
1 | 557889
2 | 001112222
2 | 5578899
3 | 1124
```

3 |
4 |
4 | 6

(b) There is no suggestion of a long-term change in the maximum rainfall at South Bend, although one very high value is recent.

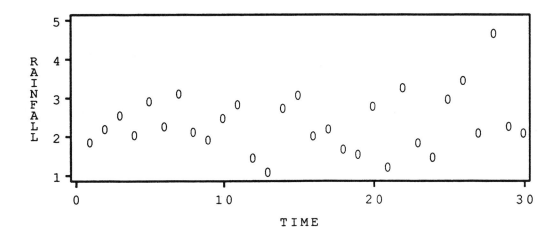

1.21 Take the coded values from Exercise 1.11, truncate the last digit and split the stems. Yes, it will have the same general appearance. The distribution is slightly skewed to the right and there is one peak.

9 | 01
9 | 66788889
10 | 01
10 | 589
11 | 2

223.9 | 01
223.9 | 66788889
224.0 | 01
224.0 | 589
224.1 | 2

1.22 (a) The distribution is skewed to the right. The values over 300 are high but appear to be part of the tail of the skewed distribution.

```
12 | 58
13 | 46
14 | 12589
15 | 66
16 | 05699
17 | 35689
18 | 012489
19 | 2268
20 | 1133
21 | 0177
22 | 0049
23 | 7
24 | 48
25 | 22
26 | 2
27 | 199
28 |
29 |
30 | 02
31 | 8
32 | 1
33 | 5
```

(b) These four cities do not have a Spring and/or Fall frost.

1.23 The classes and frequencies given below can be used to make a histogram that will be essentially the same as the stemplot. The lower endpoint is included in the interval for each class. The distribution is slightly skewed to the left.

	Group	Frequency
1 | 44	10--15	2
1 | 5899	15--20	4
2 | 2	20--25	1
2 | 55667789	25--30	8

```
3 | 13344          30--35     5
3 | 555589         35--40     6
4 | 0011234        40--45     7
4 | 5667789        45--50     7
5 | 1224           50--54     4
```

1.24 One choice is (class, freq) = (120–150, 9), (150–180, 12), (180–210, 14), (210–240, 9), (240–270, 5), (270–300, 3), (300–330, 4), (330–360, 1). The lower endpoint is included in the class.

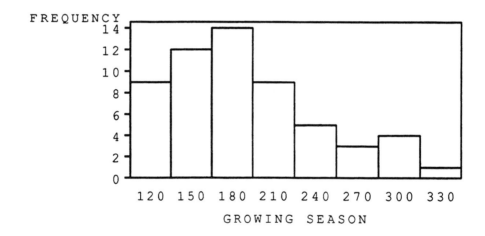

1.25 The stemplot gives more information than the histogram, but both give the same impression. The distribution is roughly symmetric with one value that is somewhat low. The center of the distribution is between 5.4 and 5.5.

```
48 | 8
49 |
50 | 7
51 | 0
52 | 6799
53 | 04469
54 | 2467
55 | 03578
```

56 | 12358
57 | 59
58 | 5

1.26 (a) The highest class is the youngest age group. The next group is relatively low. There is a smooth pattern of decrease from the third to the last group.

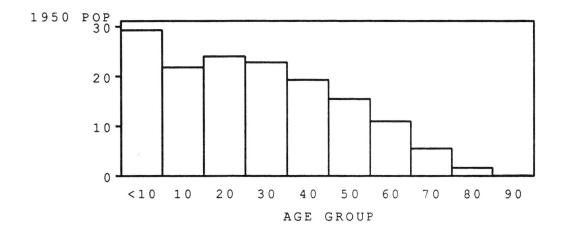

(b) The distribution is relatively flat up to the class 60–69 and then falls off. The population in 2075 will have a smaller proportion of young people and a significant number of old people.

1.27 No major differences between the two groups are evident. The centers are close together.

```
Placebo     Calcium
      8 |  9 |
      2 | 10 | 2
      9 | 10 | 77
   4421 | 11 | 0122
     97 | 11 |
      3 | 12 | 3
        | 12 | 9
      0 | 13 |
        | 13 | 6
```

1.28 If you use the first two digits for the stem and the last digit for the leaves, both distributions are very spread out and you might conclude that there are outliers. A better plot is obtained by truncating to two digits and using split stems.

```
  Control       Experimental
       87 | 2 |
    44221 | 3 | 123
 98866555 | 3 | 6799
```

CHAPTER 1

```
310 | 4 | 00001222334
 65 | 4 | 67
```

1.29 Both distributions are skewed to the right. Men tend to have higher salaries than women.

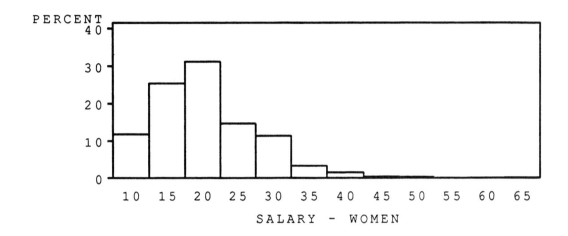

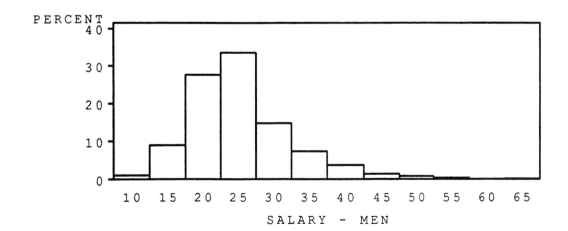

1.30 For the classes that are $10,000 wide divide the given percent by two; for the $20,000 wide class divide by four.

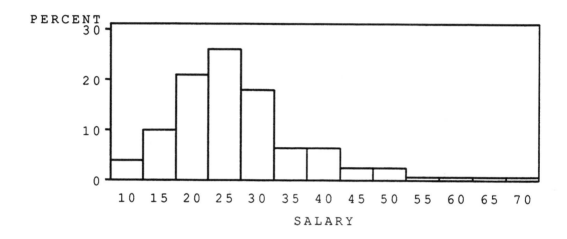

1.31 Shape is roughly symmetric. Center is around 68 inches.

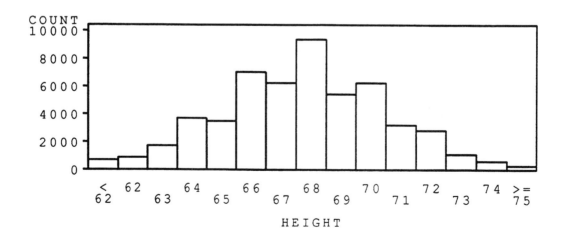

1.32 (a)

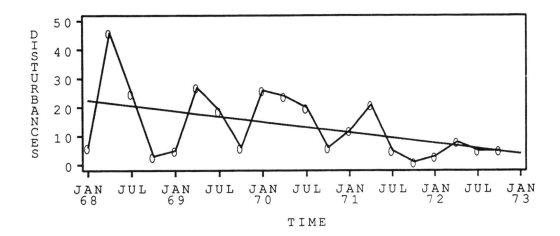

(b) The trend is decreasing. The highs appear during Apr–Jun and lows appear during Oct–Dec. The warmer weather brings more people outside.

1.33 (a) There is a general decreasing trend since 1966 with a relatively flat period from 1974 to 1980. Since 1982 the rate has been decreasing.

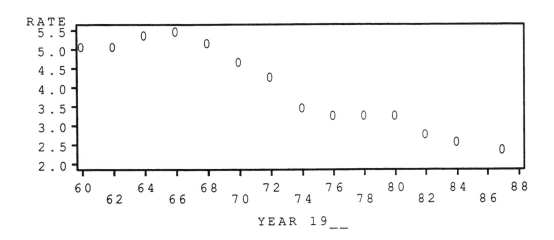

(b) The decrease starting in 1974 is evident but there is no apparent increase following the mid–eighties changes.

1.34 The first four years (1914 to 1918) are low. Starting in 1919, there is an increase and high numbers of home runs continue until a gradual decline starting around 1932. The year 1925 is unusually low.

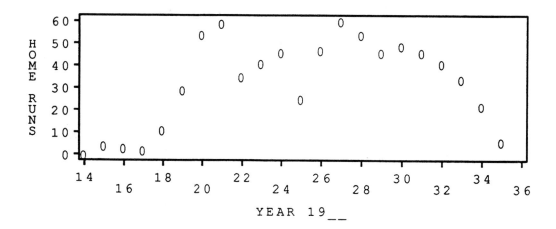

1.35 There is a general increasing trend with clear seasonal variation. There are more passengers in the summer months and fewer in the winter months.

CHAPTER 1

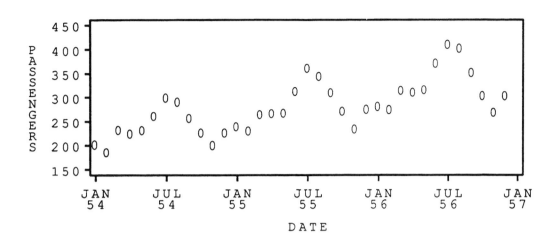

1.36 (a)

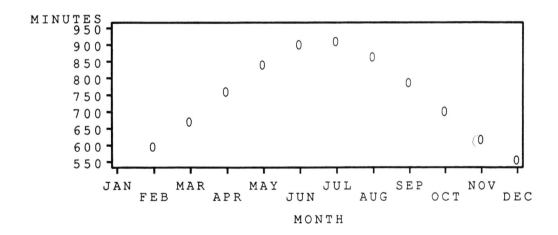

(b) Although JAN and DEC are at opposite ends of the plot, they are close in time. **(c)** 9:13, difference is 7 minutes. **(d)** 13:20, difference is 2 minutes. **(e)** 7 minutes.

1.37 (a)

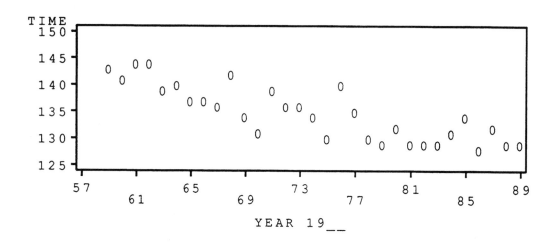

(b) Times are decreasing.

1.38 (a) The highest weights appear in the winter months, and the lowest weights are in the summer months.

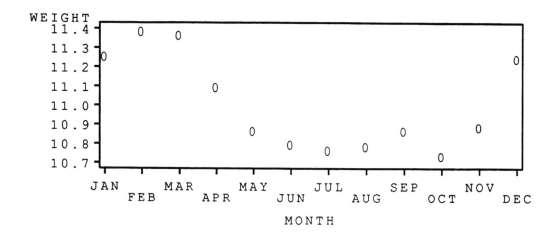

CHAPTER 1

(b) The largest heights appear in the summer months, and the smallest heights are in the winter months.

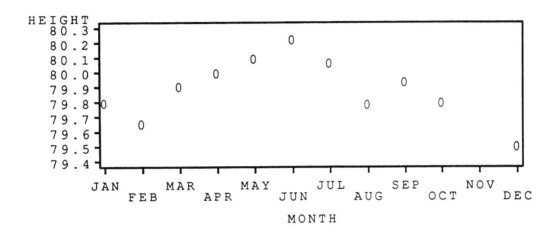

(c) The pattern for the heights is the reverse of that seen for the weights.

1.39 (a) APR, MAY, JUN, JUL are high; SEP and DEC are low.

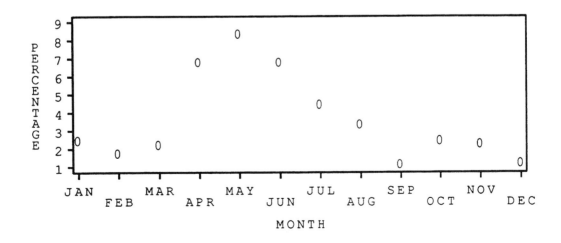

(b) Yes, the secondary peak is between OCT and NOV. **(c)** Months with high diarrhea are followed by months with low weight. Height isn't as sensitive to the short term effects of illness.

1.40 Mean = 516.3. Median = 516.5. The two measures are very close because the distribution is symmetric.

```
4 | 4
4 | 5
5 | 000244
5 | 58
```

1.41 Mean = 141. Median = 138.5. The distribution has a single outlier that causes the mean to be larger than the median.

1.42 Mean = $60,000. Seven of the eight people earn less than the mean. Median = $22,000.

1.43 Mean = 224.002. Median = 223.988. Distribution is slightly skewed to the right and the mean is a little larger than the median.

1.44 The mean increases to $113,750, while the median remains the same.

1.45 $x^* = -223000 + 1000x$. Mean = 1002. Median = 988.

1.46 We expect the distribution to be skewed to the right so the mean should be larger than the median. Meani = $159,000. i Median = $129,900.

1.47 Suppose the incomes for five people are 0, 0, 1, 2, 3 and each income increases by .5. Excluding the zeros the median decreases from 2.0 to 1.5. The mean also decreases from 2.0 to 1.7.

1.48 (a) The female scores tend to be higher than the male scores.

```
 Male     Female
  50 |  7 |
   8 |  8 |
  21 |  9 |
```

```
984  | 10 | 139
5543 | 11 | 5
   6 | 12 | 669
   2 | 13 | 77
  60 | 14 | 08
   1 | 15 | 244
   9 | 16 | 55
     | 17 | 8
  70 | 18 |
     | 19 |
     | 20 | 0
```

(b) Male: 70, 98, 114.5, 143, 187. Female: 101, 126, 138.5, 154, 200. The only outlier is the female score of 200.

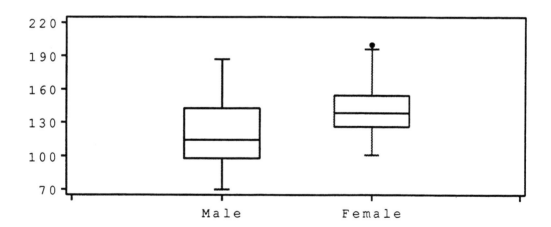

1.49 (a) 1.12, 1.88, 2.23, 2.86, 4.69. (b) IQR = .98. (c) Mean should be greater than the median since the distribution is skewed to the right.

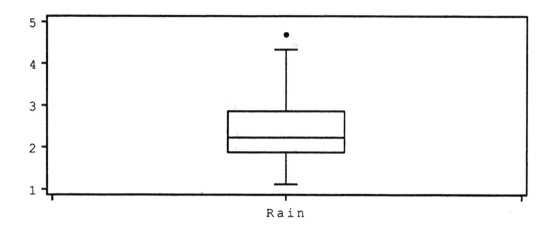

1.50 2.84, 4.78, 5.66, 7.26, 11.91.

1.51 (a) The distribution is skewed to the right.

```
 7 | 9
 8 | 136789
 9 | 015
10 | 25
```

(b) 89.67, 61.33, 7.83. **(c)** 88.5, 84.5, 93, 8.5. There are no outliers. **(d)** Quartiles, since distribution is skewed to the right.

1.52 44.17.

1.53 60, 4.97.

1.54 (a) 10, 10, 10, 10. **(b)** 0, 0, 10, 10. **(c)** Any four numbers that are all the same will have a standard deviation of 0. A proof for part (b) is beyond the scope of this text. An intuitive argument suggests that the observations should be at the extremes. Trial and error verifies that two should be at each extreme.

1.55 It is not unusual to obtain a standard deviation of 0 when there are four or more zeros in the center of the number.

1.56 5.42, .34. Because there is a clear outlier, the mean and standard deviation are not good summary statistics.

1.57 61.8, .0618 mm. There appears to be some skewness and a few gaps in the distribution. However, the mean and standard deviation are not unreasonable descriptive measures for this set of data. Either answer with an explanation is acceptable.

1.58 The mean and standard deviation are 11.14 and 2.16. Although the distribution is symmetric, there are two clear outliers. The five–number summary is preferred: 3, 9.6, 11.5, 12.4, 17.3.

1.59 Prefer median and quartiles. 435, 523, 590.

1.60 (a) 0, 1/3. (b) 0, 60. (c) -98.6, 1.

1.61 (a) 65.5. Yes. (b)

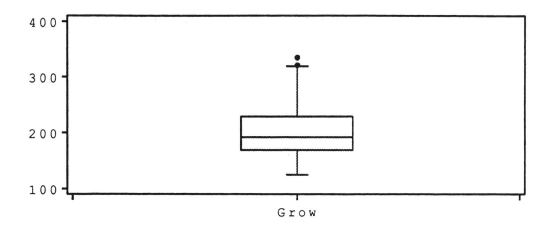

1.62 Median changes from 192 to 198; the first quartile from 167.5 to 169; the third quartile from 233 to 240.5. These changes are rather small. The mean and standard deviation cannot be computed because the growing season is undefined for these four cities.

1.63 (a) $x^* = 0 + (1/7)x$. **(b)** 29.09, 7.41, 27.43.

1.64 9.6, 12.7.

1.65 160, 181, 203, 248.

1.66 The medians are similar. Meat has a low outlier and the difference between its median and Q_1 is very small.

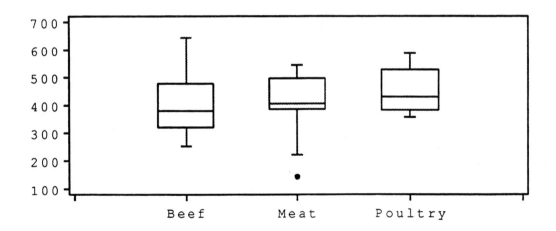

1.67 Calcium: 102, 107, 111.5, 123, 136. Placebo: 98, 109, 112, 119, 130. No.

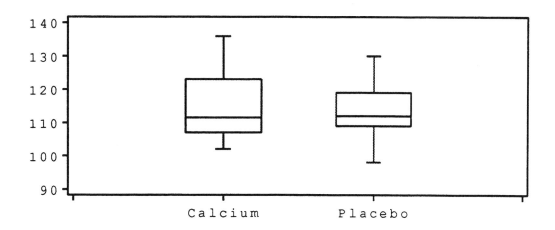

1.68 Control: 272, 337, 358, 400.5, 462. Experimental: 318, 383.5, 406.5, 428.5, 447. The median for the experimental group (406.5) is larger than the median for the control group (358).

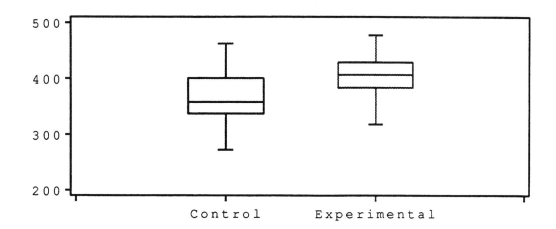

1.69 (a) $x^* = 0 + (1/28.35)x$. (b) Control: 9.59, 11.89, 12.63, 14.13, 16.30.

Experimental: 11.22, 13.53, 14.34, 15.11, 16.83. **(c)** The transformed boxplots are essentially the same.

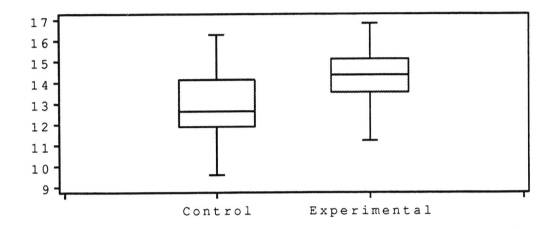

1.70 17.5, 22.5, 27.5, 32.5, 37.5.

1.71 10% trimmed mean = 2.32. 20% trimmed mean = 2.32. Median = 2.23. Mean = 2.37.

1.72 (a) C, B. (b) A, A. (c) A, B.

1.73 (a) .2. (b) .6. (c) .5.

1.74 .5, .5, .25, .75.

1.75

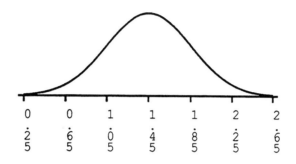

1.76

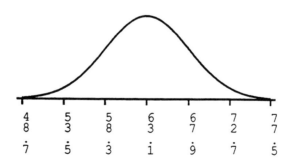

1.77 0.65 to 2.25.

1.78 (58.3, 67.9); (53.5, 72.7); (48.7, 77.5)

1.79 Eleanor: $(680 - 500)/100 = 1.8$, Gerald: $(27 - 18)/6 = 1.5$. Eleanor has the higher score.

1.80 Cobb: $(.420 - .266)/.0371 = 4.15$; Williams: $(.406 - .267)/.0326 = 4.26$; Brett: $(.390 - .261)/.0317 = 4.07$.

1.81 (a) .9978. (b) .0022. (c) .9515. (d) .9493.

1.82 (a) .0122. (b) .9878. (c) .0384. (d) .9494.

1.83 (a) –.68. (b) .25.

1.84 (a) .84. (b) .39.

1.85 (a) .7257. (b) .4514.

1.86 (a) .0655. (b) .3336. (c) .6600.

1.87 .0668.

1.88 .0516, .0776.

1.89 $Z = .52 = (x - 3)/.8$; The score is $x = .52(.8) + 3 = 3.4$. $Z = -.52 = (x - 3)/.8$; $x = (-.52)(.8) + 3 = 2.6$. Mexican Americans with scores from 1.0 (the minimum) to 2.6 are the 30% who are most Mexican/Spanish in their acculturation.

1.90 69.24.

1.91 .0301, .1995, .4295, .2815, .0594.

1.92 (a) 50%. (b) 9.18%. (c) .38%. (d) 40.82%.

1.93 124.6, 134.9.

1.94 $11/16 = 68.75\%$.

1.95 50%, 69%, 81%. Yes. The sample standard deviation found from Exercise 1.49 is .0618. Since the assumed $\sigma = .03$ is too small in this exercise, the proportions of observations within $k\sigma$ of the mean are too small.

1.96 (a) $Q_1 = -.67$, $Q_3 = .67$. (b) $IQR = .67 - (-.67) = 1.34$. (c) $1.5(IQR) = 1.5(1.34) = 2.01$. $Q_1 - 1.5(IQR) = -.67 - 2.01 = -2.68$, $P(Z < -2.68) = .0037$. $Q_3 + 1.5(IQR) = .67 + 2.01 = 2.68$, $P(Z > 2.68) = 1 - P(Z < 2.68) = 1 - .9963 = .0037$. Answer is $.0037 + .0037 = 0074 = 74\%$.

1.97 Approximately normal but the tails are less extreme than expected.

1.98 This distribution is clearly skewed to the right. The concave normal quantile plot is an indication of skewness to large values.

1.99 Approximately normal with one large outlier.

1.100 Yes the distribution is approximately normal. The vertical stacks are due to the granularity in the data. This is a consequence of the measuring instrument.

1.101 83%, 97%, 97%. Yes, there are outliers.

1.102 The normal quantile plot indicates that the data are not normal. There are at least three low outliers.

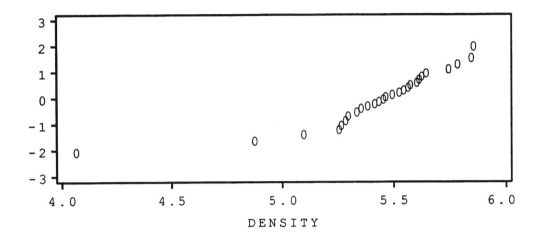

1.103 Distribution is not normal. Both tails have some extreme observations.

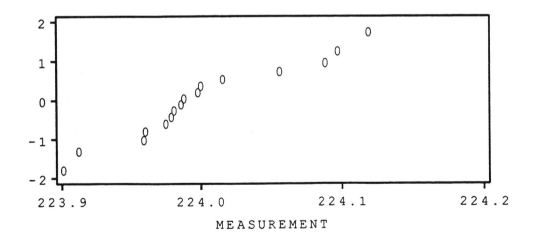

1.104 The normal quantile plots for beef and poultry hot dogs also give an indication of two distinct clusters. There is a low outlier in the beef plot and perhaps a high outlier in the poultry plot. More information about these hot dogs is needed to clarify the situation.

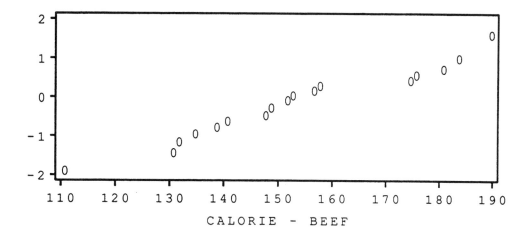

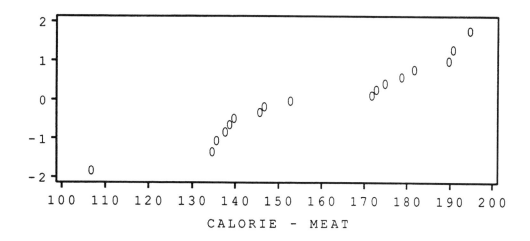

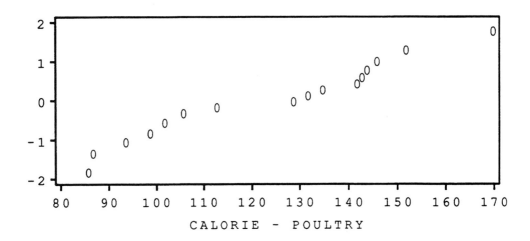

1.105 The 1940 distribution has a marked skew to the left while the 1980 distribution is less spread out and is roughly symmetric. In 1980 few states have low percentages reflecting the increased voting by blacks.

1.106 The pattern for breast cancer has not changed very much over time, whereas lung cancer is increasing. The problem is difficult because the 1985 value for lung cancer does not follow the trend present for the previous three or four points. Acceptable answers are 1990 to 2000.

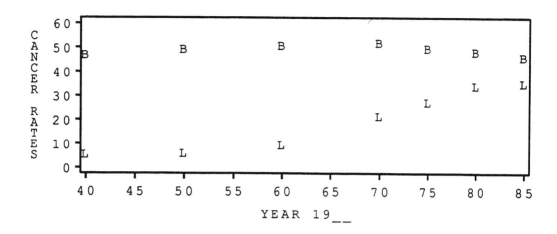

1.107 The mean and standard deviation are .9281 and .2141. The small sample size makes it difficult to assess the exact shape but there are no outliers. For the stemplot we multiply the values by 100 and round.

```
 5 | 4
 6 | 47
 7 |
 8 | 9
 9 | 048
10 | 789
11 | 3
12 | 2
```

There may be two clusters of observations but no clear deviation from normality is evident in the normal quantile plot. More data is needed to clarify.

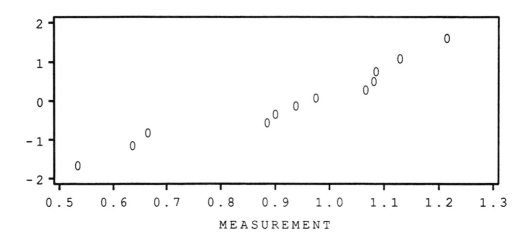

1.108 The distribution is clearly skewed toward high salaries. The mean is $949,000 while the median is $230,000. The standard deviation is $1,036,000 and is not very useful for describing this nonnormal distribution. The five-number summary is 100, 120, 230, 2100, 3300 (in thousands of dollars). Note that the four lowest salaries are $100,000, the minimum salary for major league baseball players. In the stemplot given below, the stems give millions of dollars and the leaves are rounded values of hundred thousands.

0 | 11111111222222
0 | 668
1 | 02
1 | 9
2 | 1134
2 | 66
3 | 3

1.109 The distribution is skewed to the right. The mean and standard deviation are 60.28 and 102.08. The five number–summary is 5, 19, 30, 60, 797. The IQR is 41. Possible groupings are (a) less that 50,000, (b) 50,000 to 200,000, and (c) the four largest counties.

0 | 57889
1 | 0033444567788899999

```
 2 | 0123444445566677889
 3 | 000011245557788889
 4 | 0023568
 5 | 6
 6 | 445
 7 | 246
 8 | 188
 9 |
10 | 6799
11 |
12 | 09
13 | 11
14 |
15 | 6
16 | 5
...
24 | 7
...
30 | 1
...
47 | 6
...
79 | 7
```

1.110 The mean is much higher than the median because the distribution is skewed to the right. The capitalizations of most companies are relatively small but a few are very large.

1.111 (a) $\sigma^2(x^*) = b^2 \sigma^2(x)$. $20^2 = b^2 10^2$. $b = 2$. $\mu(x^*) = a + b\mu(x)$. $100 = a + (2)(75)$. $a = -50$. **(b)** $a = -49.24$, $b = 1.82$. **(c)** $2(78) - 50 = 106$, $(1.82)(78) - 49.24 = 92.72$. **(d)** For the third graders the transformed score $X^* < 106$ corresponds to $Z < (106 - 100)/20 = .3$. The relative frequency from Table A is .6179 or 61.79%. The raw score gives the same result: $z = (78 - 75)/10 = .3$. For the sixth graders the transformed score $X^* < 92.72$ corresponds to $Z < (92.72 - 100)/20 = -.36$. The relative frequency is .3594 or 35.94%. The raw score gives the same result: $z = (78 - 82)/11 = -.36$.

1.112 (a) $X < 20$ corresponds to $Z < (20 - 25)/5 = -1.00$. The relative frequency is .1587. (b) $X < 10$ corresponds to $Z < (10 - 25)/5 = -3.00$. The relative frequency is .0013. (c) The top quarter corresponds to $z = .675$. Solving $.675 = (x - 25)/5$ gives 28.38.

1.113 (a) $a = 0$, $b = 4$. (b) $x^* = 0 + 4(30) = 120$. (c) 74.4, 83.2, 89.6, 95.0, 100.0, 105.0, 110.4, 116.8, 125.6.

1.114 The normal quantile plot indicates that the data are approximately normally distributed. The mean and standard deviation are good measures for normal distributions. Mean = 35.09. Standard deviation = 11.19.

1.115 $x^* = 68.6393 + .8937x$.

1.116 (a)

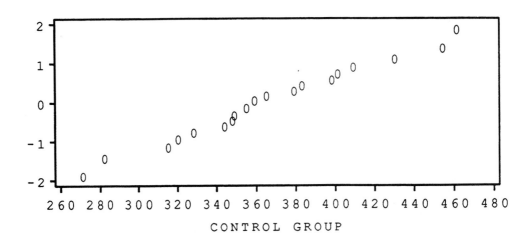

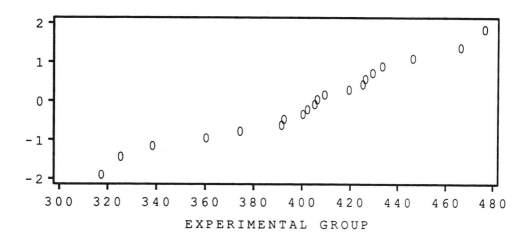

(b) (Mean, Standard deviation) = 366.3, 50.81) for control; (402.95, 42.73) for experimental. The high–lysine chicks gained more weight. The standard deviations in the two groups are similar.

1.117 (a) There are high values that lie to the right of a line drawn through the main body of points indicating that the distribution is strongly skewed to the right.

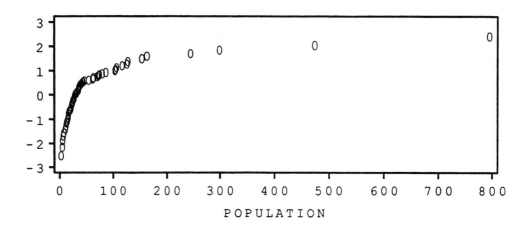

(b) The mean and median are 60.28 and 30. They differ because the distribution is skewed to the right. The total can be recovered only from the mean; it is 5546. **(c)** The counts are 87, 89, and 90. These counts correspond to 95%, 97% and 98%. The rule does not apply.

1.118 (a)

```
186 | 0
187 | 26
188 |
189 |
190 | 224
191 | 8
192 | 25
193 | 3456777
194 | 23478
195 |
196 | 0179
197 | 3344567899
198 | 00113
```

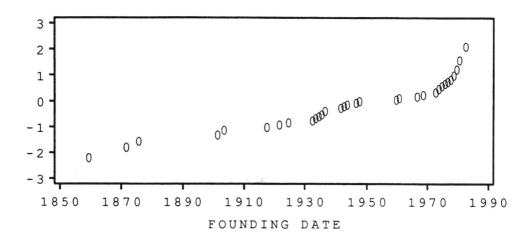

(b) 1946.85, 32.99. (c) Five-number Summary: 1860, 1933.5, 1947.5, 1975.5, 1983. IQR = 42. (d) The distribution is clearly skewed toward low values. There are three clusters of founding dates. The first group of three were founded before 1900. The next group were founded between 1900 and 1948. The distribution of this group is not symmetric with more observations in the 30's and 40's. The last group are relatively new wineries founded since 1960. The mean and standard deviation are not particularly useful. The five-number summary does not adequately describe this set of data.

1.119 (a)

```
14 | 5
15 | 055778
16 | 000235558
17 | 015555689
18 | 00000123344555556889
19 | 0023555566
20 | 00000258
21 | 0000444455558888
22 | 000024455555889
23 | 0034555
24 | 15
```

```
25 | 02356
26 | 00000033556
27 | 00666
28 | 02
29 | 00445
```

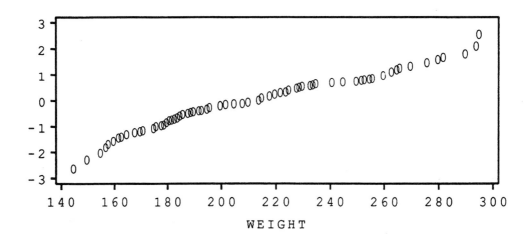

(b) It is not symmetric, not normal. There are three peaks that probably correspond to different positions of the players. There are no outliers. **(c)** The 5–number summary is 145, 183, 210, 235, 295. There are peaks at approximately 185, 220, 260. **(d)** The 5–number summary is 65.9, 83.2, 95.5, 106.8, 134.1. The peaks are 84.1, 100.0, 118.2. **(e)** Many digits end in zero or five.

1.120 The following stemplot was generated by SAS. The data were rounded off to two digits and the stems were split. Note that this program has rounded 145 to 150 and has placed this on the first 1 stem as a 5 rather than on the second 1 stem. The distribution is skewed to the right. Five–number Summary – 43, 82.5, 102.5, 151.5, 598.

```
0 | 4
0 | 5566667777888888888999999
```

```
1 | 000000000000111112223344445
1 | 56678889
2 | 0114
2 | 5
3 | 3
3 | 8
4 | 0
4 |
5 | 12
5 |
6 | 0
```

1.121 Code for MINITAB is given in the problem. For SAS the following can be used.

```
Data; Do i=1 to 100; x=20+5*rannor(0); Output; End;
Proc Univariate; Var x;
Run;
```

The distribution of the means looks normal with mean 20.07 and standard deviation 1.12. The standard deviations have a high peak in the middle of the distribution; the shape is approximately symmetric; the mean and median are 4.98 and 5.01. Results will vary with different simulations. This is a stemplot of the 20 values of $\bar{x}$.

```
18 | 346
19 | 13355
20 | 00224457
21 | 08
22 | 02
```

This is a stemplot of the 20 values of s.

```
3 | 8
4 | 023
4 | 669
5 | 0000112334
5 | 6
6 | 01
```

1.122 (a) For the salaries as given, the five-number summary is 100, 120, 230, 2100, 3300 and the IQR is 1980. The mean is 949 and the standard deviation is 1036. For the new variable, the five-number summary is 0, 20, 130, 2000, 3200 and the IQR is 1980. The mean is 849 and the standard deviation is 1036. The standard deviation and the IQR are the same. For all other measures the values for the new variable are 100 less than the values for the salaries. **(b)** The shape has not changed because we simply subtracted 100 from each observation.

1.123 For H2S, $\bar{x} = 5.942$, $s = 2.127$, the median is 5.329, and the IQR is 3.689. The distribution is skewed toward high values.

```
 3 | 01278899
 4 | 27899
 5 | 024
 6 | 1278
 7 | 0569
 8 | 07
 9 | 126
10 | 2
```

For LACTIC, $\bar{x} = 1.442$, $s = 0.303$, the median is 1.45, and the IQR is 0.43. The distribution is approximately normal.

```
 8 | 6
 9 | 9
10 | 689
11 | 56
12 | 5599
13 | 013
14 | 467
15 | 2378
16 | 38
17 | 248
18 | 1
19 | 09
20 | 1
```

1.124 The distribution of SAT math scores is slightly higher for men than for women. The grade point means and medians are very close for the two groups but both quartiles are a little higher for women. All four distributions are approximately normal.

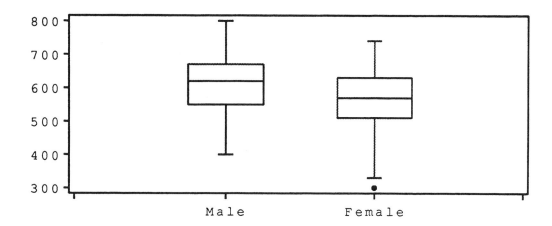

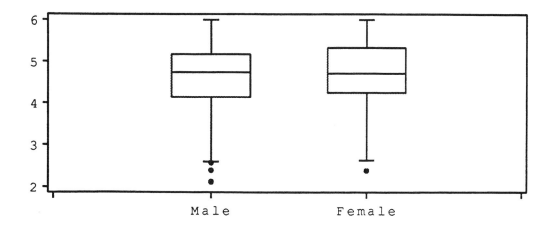

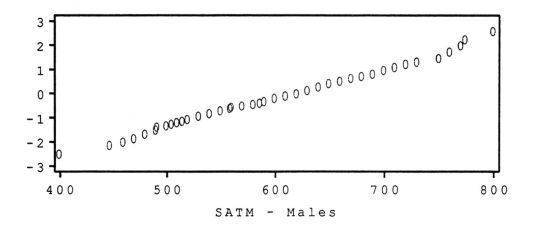

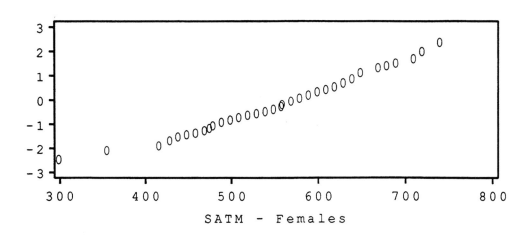

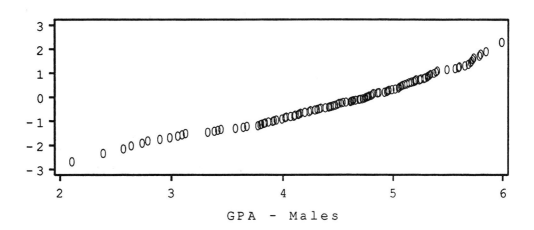

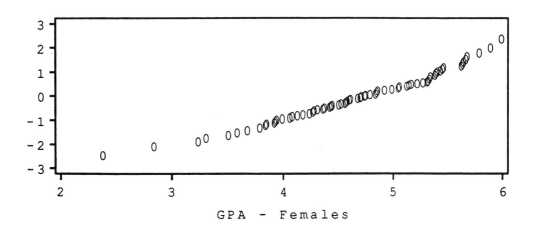

4.2 CHAPTER 2

2.1 (a) Categorical. **(b)** Quantitative. **(c)** Quantitative. **(d)** Categorical. **(e)** Quantitative. **(f)** Quantitative.

2.2 (a) The explanatory variable is time spent studying and the response variable is the grade. **(b)** Explore the relationship. **(c)** The explanatory variable is rainfall and the response variable is yield. **(d)** Explore the relationship. **(e)** The explanatory variable is the father's occupational class and the response variable is the son's.

2.3 (a) The two variables are negatively related. It is clearly curved. One observation is high on nitrogen oxides. **(b)** No. Low nitrogen oxide is associated with high carbon monoxide.

2.4 (a)

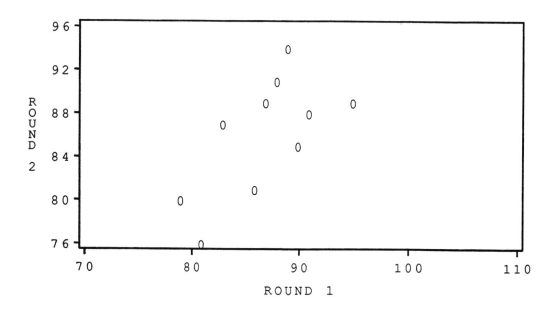

(b) There is a positive association between two scores. We would expect this because differences in skill cause some people to score well on both rounds and others to score poorly. **(c)** We cannot tell whether the value is from a good player or a poor player because there is one good round and one poor

CHAPTER 2

round. More scores on this player would be needed.

2.5 (a) Flow rate is the explanatory variable.

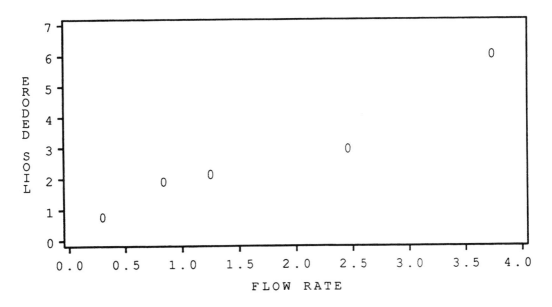

(b) As the flow rate increases the amount of eroded soil increases. Yes, the pattern is approximately linear. The association is positive.

2.6 (a)

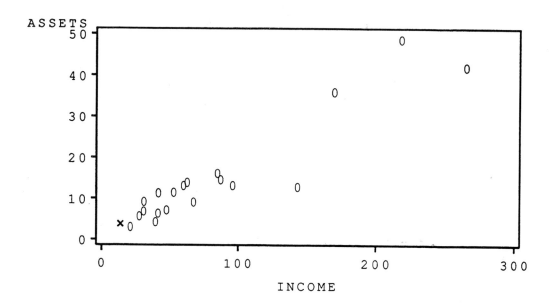

(b) There appears to be a linear association between assets and income. There a few banks with high values of both but the fit the pattern. Franklin is low on both but does not stand out.

2.7 (a) Heavier cars cost more. The association is weak and positive. **(b)** Foreign cars generally cost more than domestic cars of similar weight. The five lightest cars are foreign.

2.8 (a) Lean body mass is the explanatory variable.

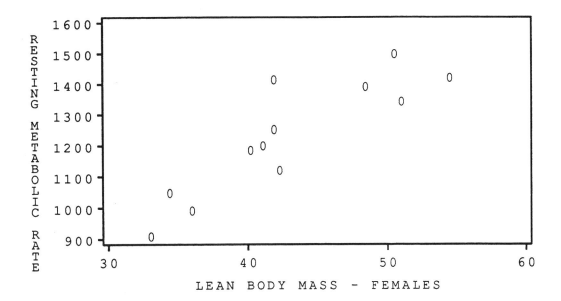

(b) The association is positive. The relationship is approximately linear. **(c)** The same linear relationship appears to hold for men. The values for men tend to be higher for both variables.

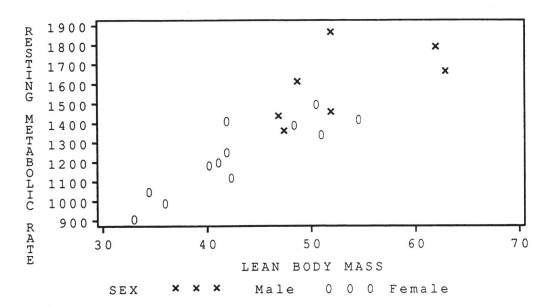

2.9 (a) Alaska is an outlier only in median teacher salary. **(b)** There is no apparent relationship between teachers' salaries and students' SAT scores.

2.10 (a)

(b) No clear relationship is evident. **(c)** Those who survive appear to be older and have longer incubation periods. **(d)** The two survivors with incubation periods greater than 70 should be investigated.

2.11 The relationship is negative and not linear.

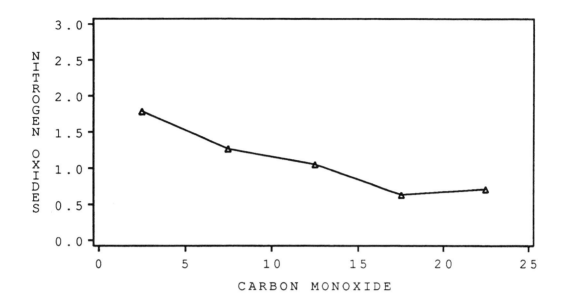

2.12 The number of fleas increases and then decreases.

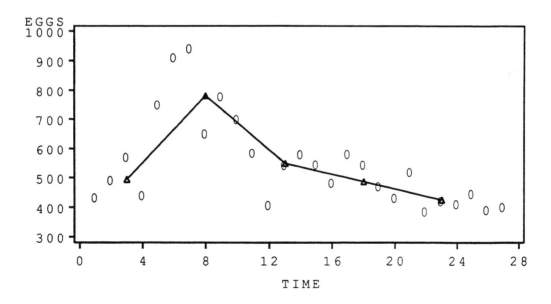

2.13 (a) Means = 10.65, 10.43, 5.60, 5.45.

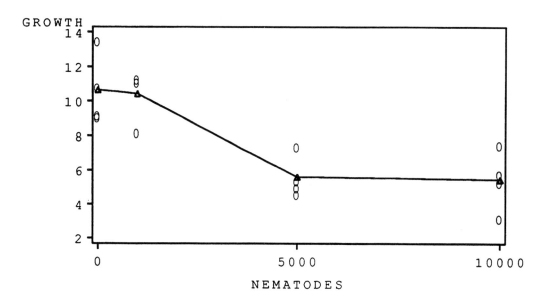

(b) The introduction of 1000 nematodes per pot has no effect on seedling growth. With 5000 nematodes there is a substantial reduction in seedling growth. Introduction of 10000 nematodes causes essentially the same growth reduction as 5000.

2.14 (a) Means = 47.17, 15.67, 31.50, 14.83.

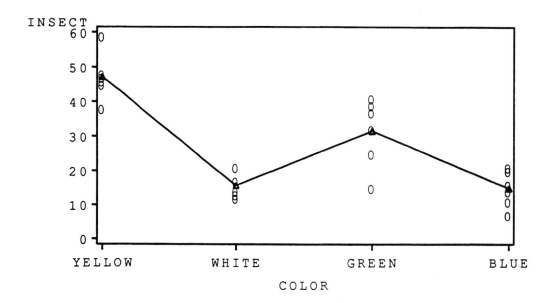

(b) Beetles are most attracted to lemon yellow. The next most attractive color is green. White and blue are the least attractive colors. (c) No. Color is a categorical variable.

2.15 (a) Means = 1520, 1707, 1540, 1816.

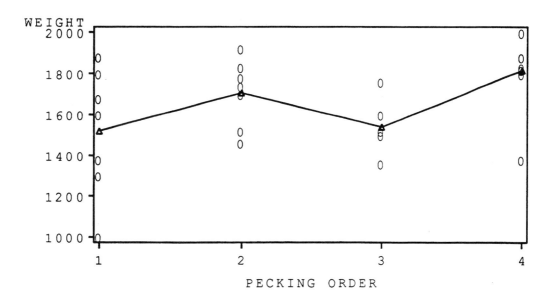

(b) Against. Pecking order 1 has the lowest mean weight, and pecking order 4 has the highest mean weight.

2.16 (a)

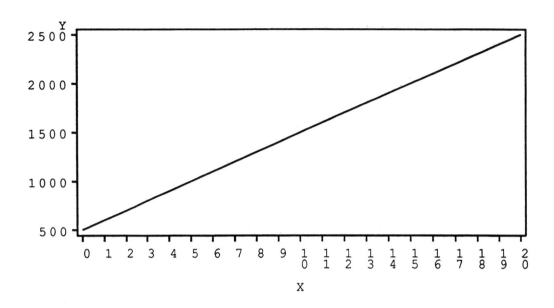

(b) $y = 500 + (100)20 = 2500.00$. **(c)** $y = 500 + 200t$.

2.17 $y = 1500x$.

2.18 (a) $w = 100 + 40t$. Slope=40. **(b)**

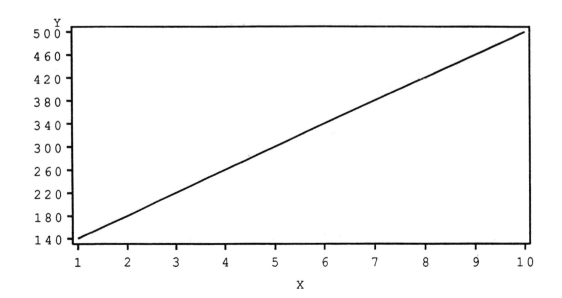

(c) No, predicted weight would be 9.4 lbs.

2.19 $y = 30+33t$. $y = 30+33(15) = 525$. $z = 50+25t$. $z = 50+25(15) = 425$. The second company is cheaper.

2.20 (a)

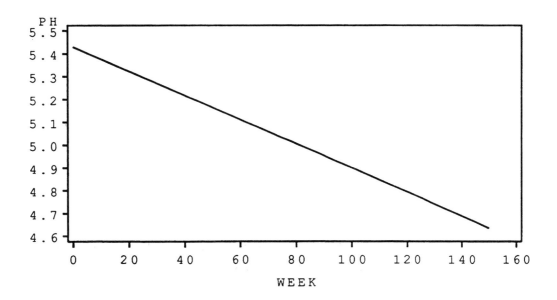

(b) $pH = 5.43 - (0.0053)0 = 5.43$. $pH = 5.43 - (0.0053)150 = 4.64$. (c) The slope is -0.0053. The pH is decreasing.

2.21 (a) (b)

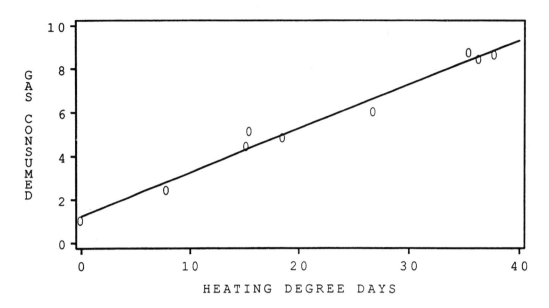

(c) 9.32. The insulation reduced gas consumption.

2.22 (a) $b = 1.389$, $a = .406$. **(b)** .836, 1.586, 2.156, 3.837, 5.615. **(c)** −.016, .364, .024, −.827, .455. These residuals sum to zero with no roundoff error. **(d)** The pattern is random. There is no evidence to suggest that a curve would give a better fit.

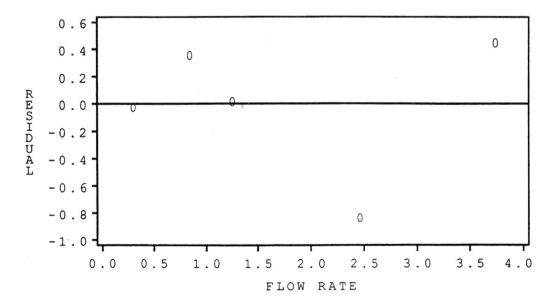

2.23 (a)

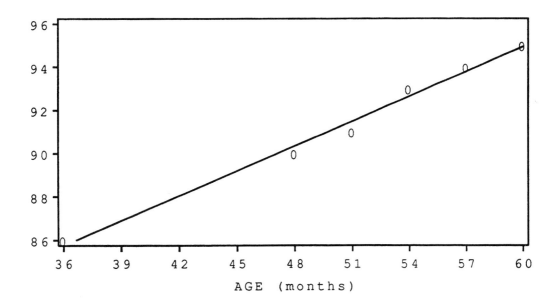

(b) $\hat{y} = 71.95 + .38x$. **(c)** Sarah grew .38 cm per month. Normal growth is .5 cm per month. Sarah's growth is less rapid than normal. **(d)** $71.95 + .38(40) = 87.15$. $71.95 + .38(65) = 96.65$.

2.24 (a) $\hat{y} = -.435 + .214x$. **(b)**

(c) $\hat{y} = -.435 + .214(4.89) = .61$.

2.25 Predicted Height = 85.75, 90.35, 91.50, 92.65, 93.80, 94.95. Residuals = 0.25, −0.35, −0.50, 0.35, 0.20, 0.05. There is no clear pattern in the residuals.

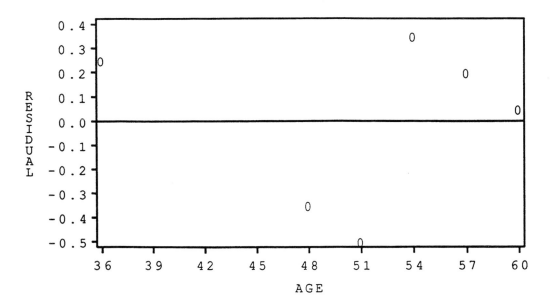

2.26 Residuals: −.050, −.058, −.048, .076, .080. These residuals sum to zero with no roundoff error. There are two high residuals and three low residuals. No pattern is evident.

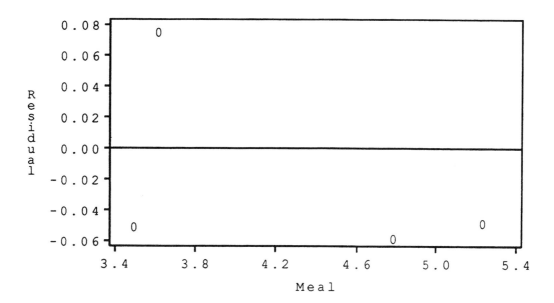

2.27 (a) Yes. The straight line fits these data well.

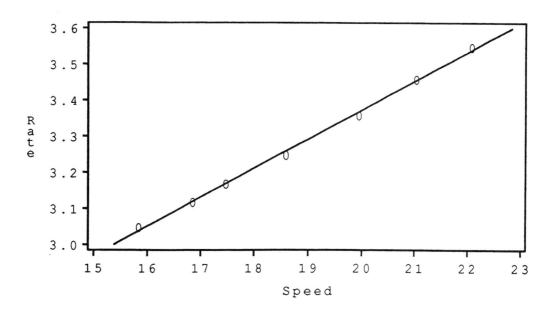

(b) $b = .080$, $a = 1.766$. (c) The predicted values are 3.039, 3.121, 3.171, 3.261, 3.369, 3.457, 3.541. The residuals are .011, −.001, −.001, −.011, −.009, .003, .009. These residuals sum to .001. The difference is due to roundoff error. (d) The residuals are all very small, indicating that the line fits the data well. Positive residuals are associated with high and low speeds and negative residuals are associated with intermediate speeds. Although a line fits the data very well, the plot of the residuals indicates that there is some curvature in the relationship. We cannot plot the residuals versus the time at which the observations were made because this information is not given.

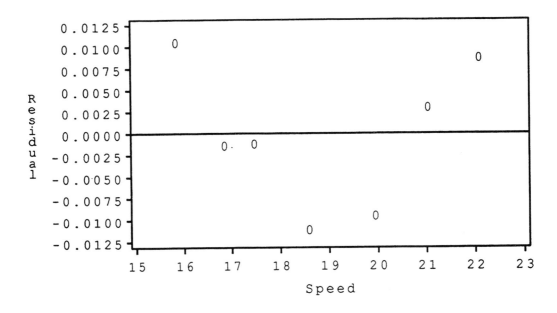

2.28 (a)

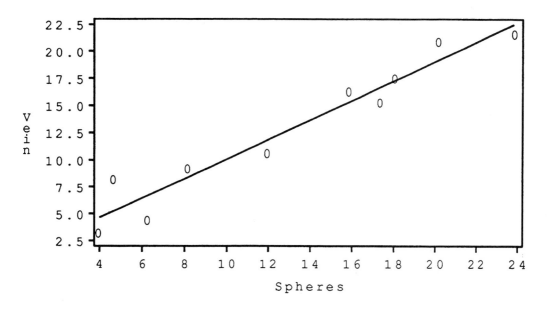

(b) $\hat{y} = 1.03 + .90x$. **(c)** Predicted = 6.43, 11.83, 17.23. Yes, the actual values are within 10% of the predicted values.

2.29 (a) The last observation is most influential because the cost is very high and the tip is zero.

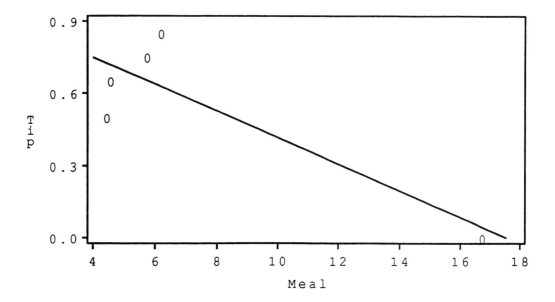

(b) The line does not fit the pattern. (c) The residual for the influential observation is the smallest of the five. This observation is so influential that it forces the line to fit it well.

2.30 (a)

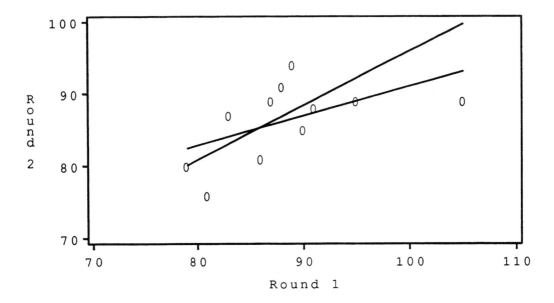

(b) The line that omits the influential observation has a steeper slope and gives the influential observation a large residual.

2.31 (a)

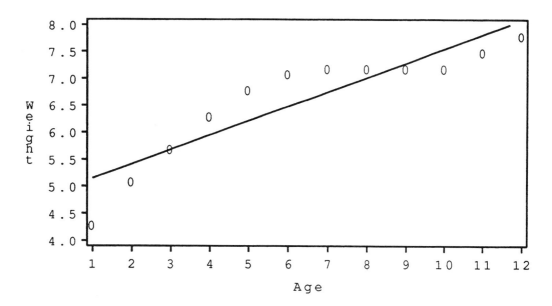

(b) There are systematic deviations from the pattern of linear growth in this plot. The line is not an acceptable summary. **(c)** The third residual is .02 not .18. The sum is .01; the difference is due to roundoff error. Residuals are increasing from 1 to 5 months and decreasing from 6 to 10 months.

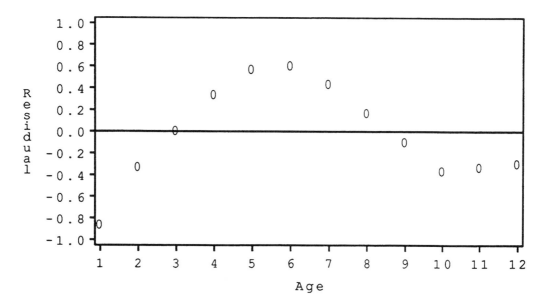

2.32 (a) Rural = 108, City = 123. No, this observation does not have the largest residual. (b) 7.4820. (c) Yes, it appears that a line will fit these data well. (d) $\hat{y} = -2.580 + 1.0935(88) = 93.65$.

2.33 (a) The relationship is linear with two clusters and one observation that is low in calories and sodium. (b) Colored. It is closer to the influential observation. (c) $\hat{y} = 46.90 + 2.401(150) = 407.05$.

2.34 (a) The residuals are more scattered for higher predicted values. The regression model will predict low salaries more precisely than high salaries. (b) The residuals are clearly related to the number of years. They are high for the middle numbers of years and low otherwise. The model will overestimate the salaries of new players. It will underestimate the salaries of players in the major leagues about eight years. It will overestimate the salaries of players with more than fifteen years in the majors.

2.35 (a)

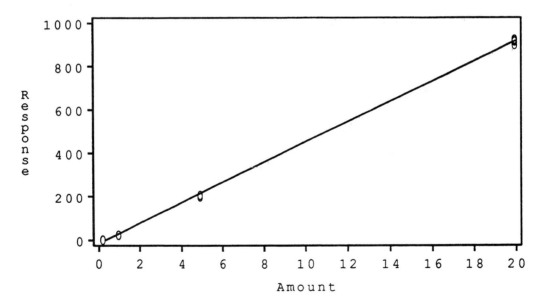

(b) $\hat{y} = -14.4107 + 46.6287x$. (c) The nonlinearity is clearly evident. The residuals for $x = .25$ are all positive; the residuals for the next two amounts are all negative. The residuals for $x = 20.0$ show large variation.

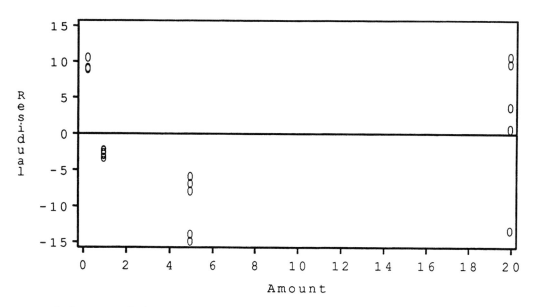

2.36 $2^{1\times 4} = 16$, $2^{5\times 4} = 1,048,576$.

2.37 (a) 1, 2, 4, 8, 16, 32, 64, 128, 256, 512. **(b)**

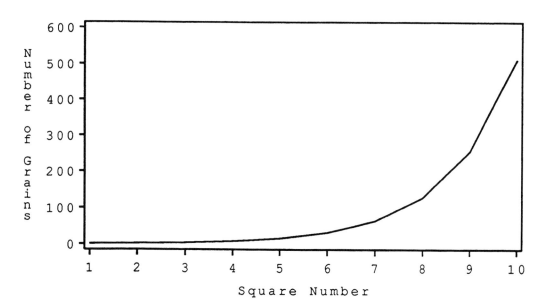

(c) Approximately 9,000,000,000,000,000,000. **(d)** 0.00, 0.30, 0.60, 0.90, 1.20, 1.51, 1.81, 2.11, 2.41, 2.71.

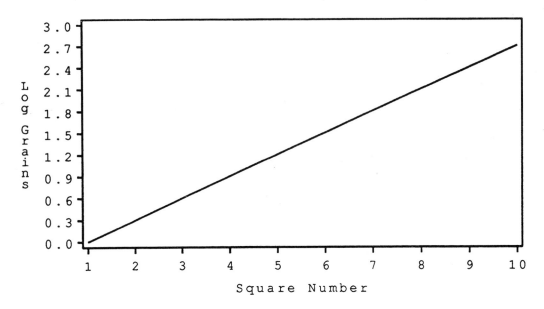

(e) $b = .3$, $a = -.3$, $-.3 + .3(64) = 18.9$. The log of the answer in part (c) is 18.95.

2.38 (a) $y = 500(1.075)^{year}$. 537.50, 577.81, 621.15, 667.73, 717.81, 771.65, 829.52, 891.74, 958.62, 1030.52. **(b)**

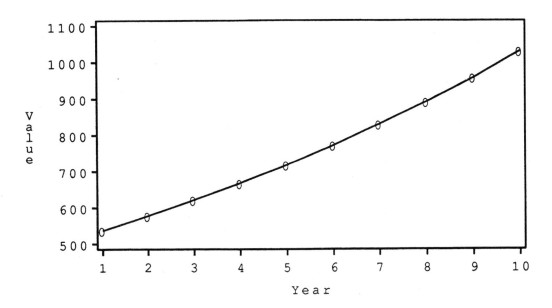

CHAPTER 2

(c) The logs are 2.73, 2.76, 2.79, 2.82, 2.86, 2.89, 2.92, 2.95, 2.98, 3.01.

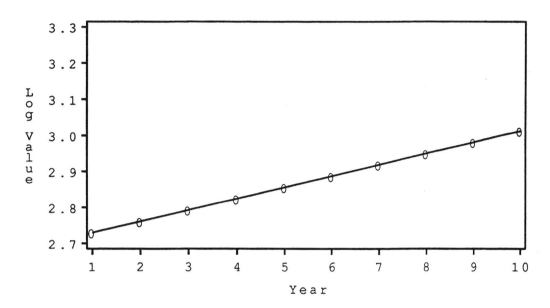

2.39 Alice has $500(1.075)^{25} = 3049.17$, Fred has $500 + 100(25) = 3000.00$.

2.40 (a)

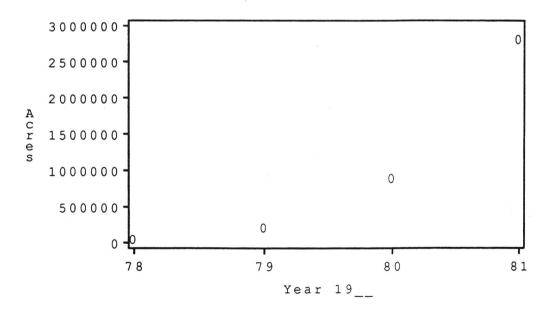

(b) The ratios are 3.59, 4.01, 3.12. **(c)**

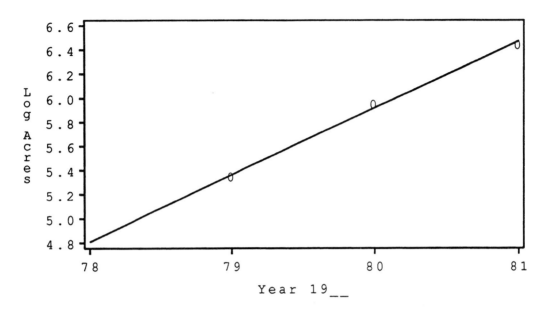

(d) The predicted value is 7.026. $10^{7.026} = 10,616,956$.

2.41 (a) The pattern appears to be exponential.

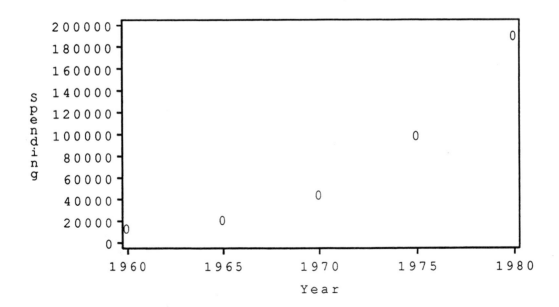

(b) Yes.

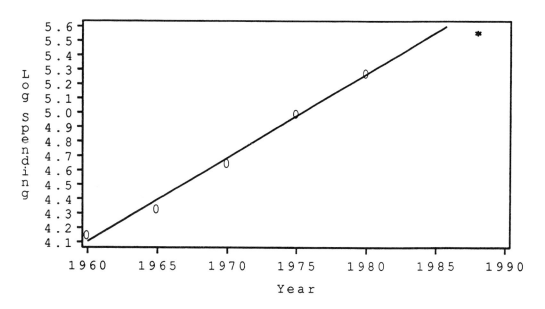

(d) Predicted log for 1988 is 5.74112, predicted amount spent in 1988 is $550,959.91. (e) Log is 5.55. The point is somewhat below the line. The rate of growth appears to be slower.

2.42 (a)

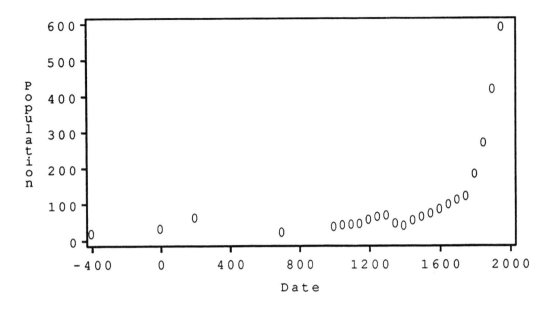

(b)

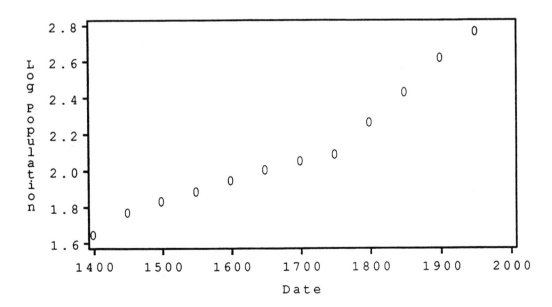

(c) The growth between 1400 and 1950 cannot be described by one exponential function. The growth from 1400 to 1750 was exponential with one rate, and the growth from 1750 to 1950 was exponential with another rate.

2.43 (a)

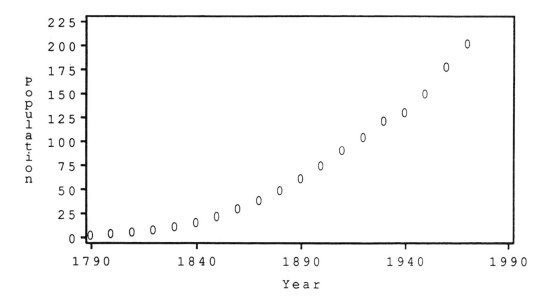

(b) The log plot looks linear from 1790 to 1880 and also from 1880 to 1990. The slope is less in the second period.

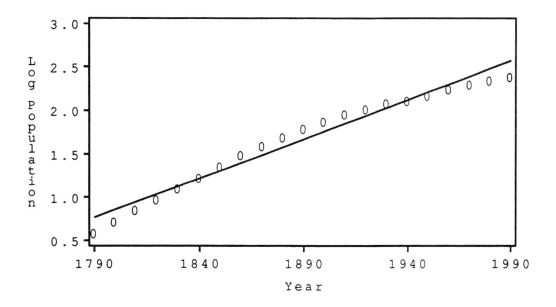

(c) Log is 2.4695, predicted population is 294.8 million.

2.44 (a)

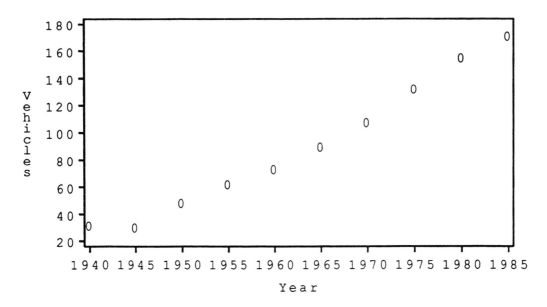

(b) The logs are 1.51, 1.49, 1.69, 1.80, 1.87, 1.96, 2.04, 2.12, 2.19, 2.23.

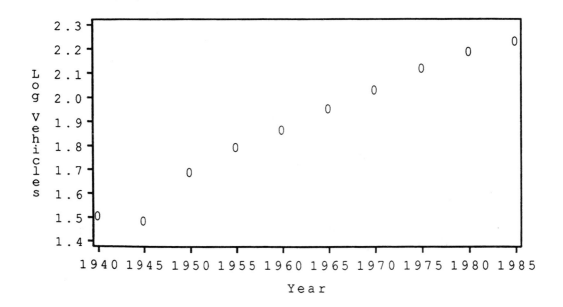

(c) The plot with the logs looks more linear in the period 1950 to 1980. Therefore, the growth is approximately exponential. (d) 1945 corresponds to the end of World War II. The pattern of exponential growth appears to be ending in 1985. (e) The predicted log is 2.349. The predicted number of motor vehicles is 223.385 million. The extrapolation overpredicted.

2.45 (a) Predicted logs are: 1.620 and 2.349. Predicted motor vehicles are: 41.7 and 223.4. **(b)** No.

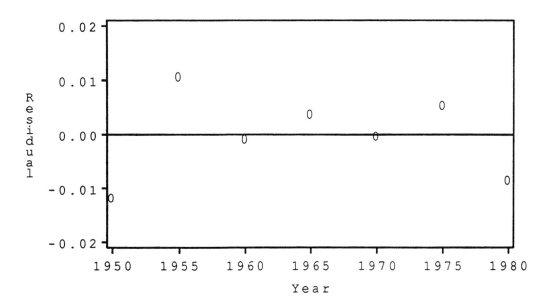

(c) The residuals for 1945 and 1989 are both very low.

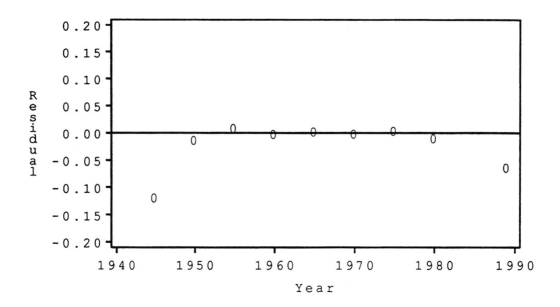

2.46 There is some evidence to support exponential growth.

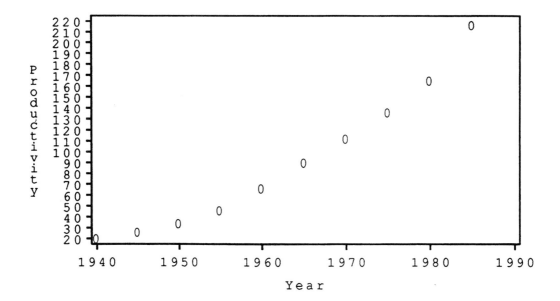

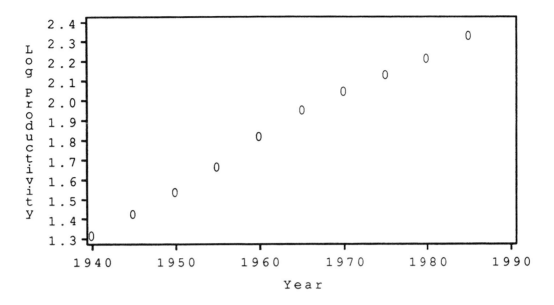

2.47 (a)

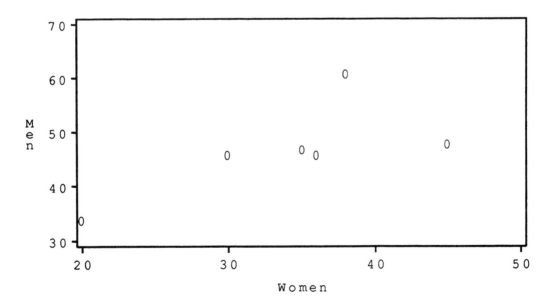

(b) There is a positive association; it is approximately linear. Bowdin produces an unusually high percent of female doctorates relative to the percent of male doctorates in science. (c) $r = .69542$. $r^2 = .48$

2.48 (a) We expect the correlation to be positive. The relationship is weak so we do not expect it to be near 1.

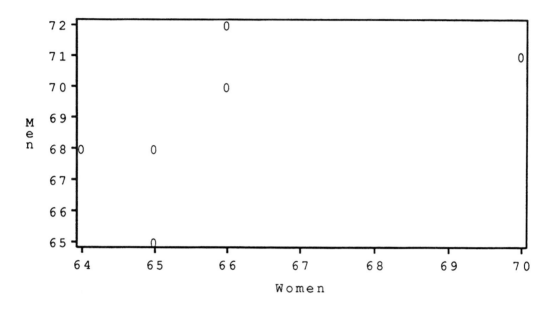

(b) $r = .56533$. **(c)** It does not change. No, we would need to examine the differences in the heights. **(d)** $r = 1$.

2.49 (a) All the points fall very close to a line. **(b)** $r = .99438$. **(c)** The value of r does not change.

2.50 Means = 69, 66; standard deviations = 2.52982, 2.09762. Slopes = .6818, .4687. At the means.

2.51 Means = 91.5, 51.0; standard deviations = 3.27, 8.49. $b = .99438(3.27109)/8.48528 = .38$. $b = .99438(8.48528)/3.27109 = 2.58$.

2.52 The correlation is .66. Player 7's scores do not fit the linear pattern of the other data. By removing this point, the strength of the linear association, as measured by the correlation, is therefore increased. See Exercise 2.14 for the plot.

2.53 (a)

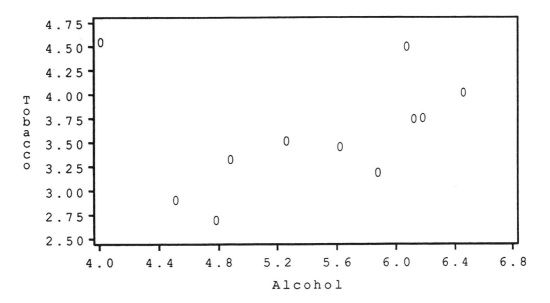

(b) The relationship is positive and linear with one outlier. **(c)** $r = .784$. The outlier almost destroys the linear relationship.

2.54 $\sum xy - (1/n)(\sum x)(\sum y) = 5360 - (1/5)(200)(134) = 0$. The association is not linear.

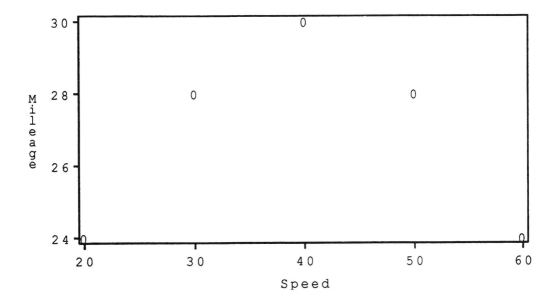

2.55 The paper suggests a negative relationship. The psychologist is saying there is no linear relationship.

2.56 (a) As we have used correlation, it applies only to quantitative variables. **(b)** The correlation must be between -1 and $+1$. **(c)** A correlation has no unit of measurement.

2.57 $r = \sqrt{.16} = .4$.

2.58 (a) $r = .56533$. The correlation does not depend on the units of measurement. **(b)** Let x be the female height in inches, y be the male height in inches and y^* be the male height in centimeters. From Exercise 2.50, $s_y = 2.52982$, $s_x = 2.09762$, $\bar{x} = 66$, $\bar{y} = 69$. $s_{y^*} = 2.54(2.52982) = 6.4257$. $b^* = rs_{y^*}/s_x = .56533(6.4257)/2.09762 = 1.7318$. $\bar{y}^* = 2.54(69) = 175.26$. $a^* = \bar{y}^* - b^*\bar{x} = 175.26 - 1.7318(66) = 60.9612$. Note that if the equation is found in inches first, the coefficients can be multiplied by 2.54 to obtain this answer.

2.59 (a) The problem asks for a plot of x^* against y^*. This could be interpreted as meaning that x^* should be plotted on the y axis and y^* on the x axis. The plot below uses the natural correspondence. Either version is acceptable.

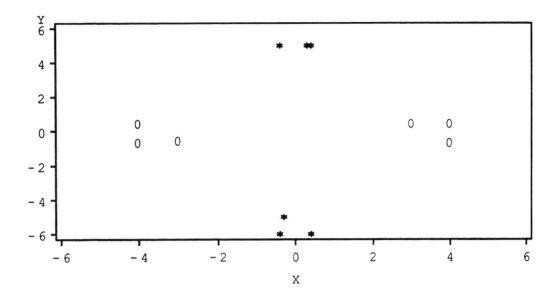

(b) $r = .25$. (c) No. It will be 100 times as large.

2.60 The correlation would be higher because the points would be more tightly clustered around a line.

2.61 It is the percentage of variation in the number of students enrolled in 100 level math courses explained by the linear relationship with the number of students in the freshman class.

2.62 93.5%.

2.63 $b = rs_y/s_x = .5(2.7)/(2.5) = .54$, $a = \bar{y} - b\bar{x} = 68.5 - .54(65.5) = 33.13$. $\hat{y} = 33.13 + .54(67) = 69.31$.

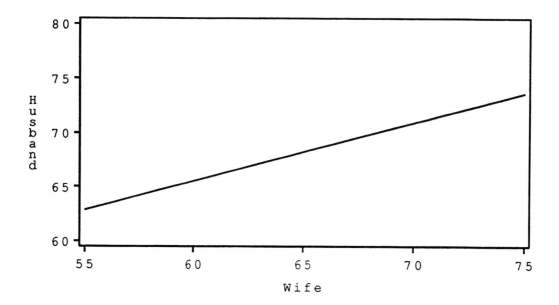

2.64 (a) $b = rs_y/s_x = .16$. **(b)** $a = \bar{y} - b\bar{x} = 30.2$. $\hat{y} = 30.2 + .16(300) = 78.2$.

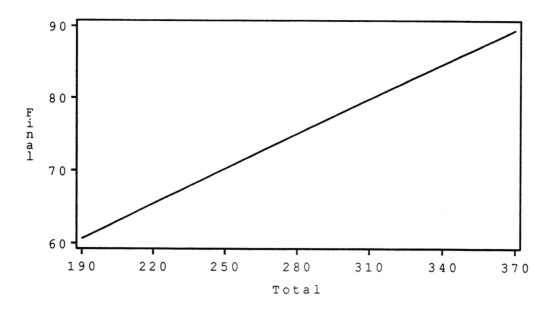

2.65 (a)

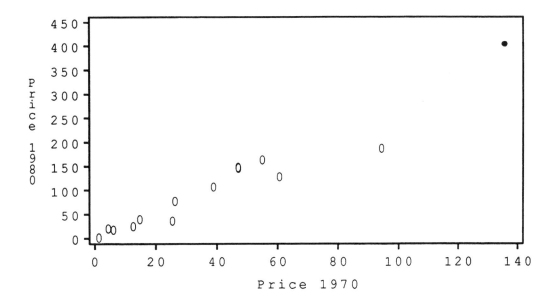

(b) There is a positive linear relationship. Sea scallops has high values for both variables; lobster is also somewhat high on both. We have chosen to call only sea scallops extreme, although an answer including both is acceptable. There are no points that are outliers in the regression sense of having large residuals from a fitted line. There are no influential observations that would substantially change the fit. (c) .96704. (d) .9352. (e) .93996. The correlation is reduced by a small amount. The effect is not large because the discarded point fits the linear pattern of the other points. (f) The correlation indicates that the linear relationship is strong.

2.66 $r = 1.0024$. The answer is not reasonable because a correlation cannot be larger than one. The roundoff has destroyed the accuracy of our calculations.

2.67

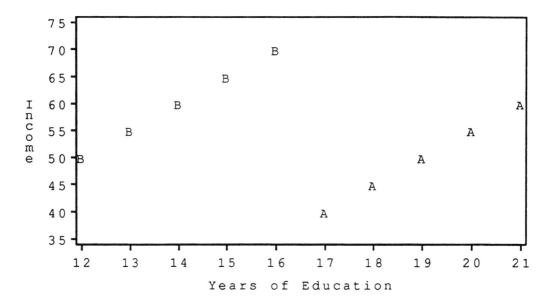

2.68 Observations for individuals would be more variable than averages. The correlation should be lower. The following plot is an example of a reasonable answer. Many variations are acceptable as long as they have more variation than the plots of means given in Exercise 2.27.

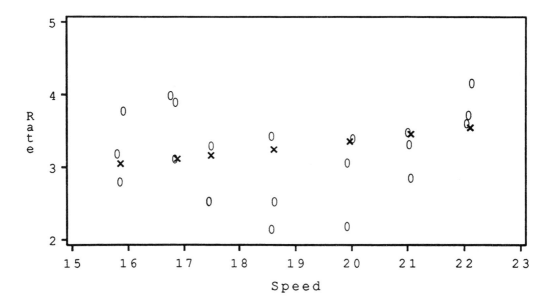

2.69 (a) $246/600 = .41$. (b) Small: $125/200 = .625$; Medium: $.405$; Large:

.200. (c)

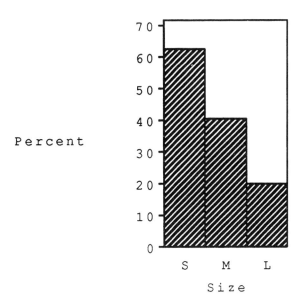

(d) Small: 125/246 = .5081; Medium: .3293; Large: .1626. (e) The small businesses are overrepresented and the large businesses are underrepresented compared to what was designed for the study.

2.70 (a) 45.90%, 45.08%, 9.02%. (b) The conditional percentages for Egg are 45.95, 44.59, 9.46. The conditional percentages for Turkey are 45.83, 45.83, 8.33. (c)

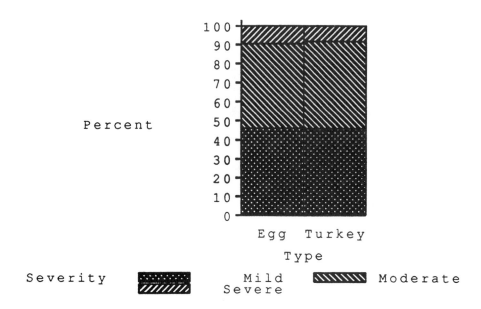

(d) Yes. The percentages for mild, moderate and severe are similar for each type of operation.

2.71 (a)

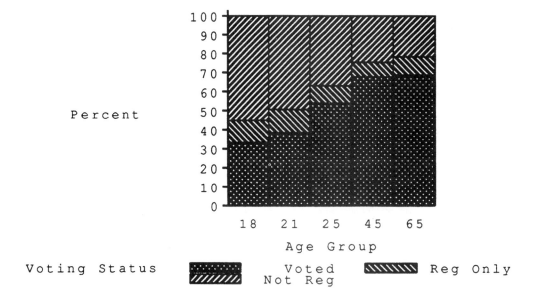

(b) The proportion of people voting increases with age. **(c)** No, you would need the total number of registered voters in each age group.

2.72 (a) 21974. Roundoff error. (b) 53.10%, 32.84%, 8.69%, 5.37%. (c) 4 yr H.S.: 26.47%, 40.63%, 18.90%, 14.00%. 1-3 yr Col: 18.37%, 36.82%, 22.60%, 22.20%. $\geq$ 4 yr Col: 8.40%, 27.30%, 22.80%, 41.50%. (d)

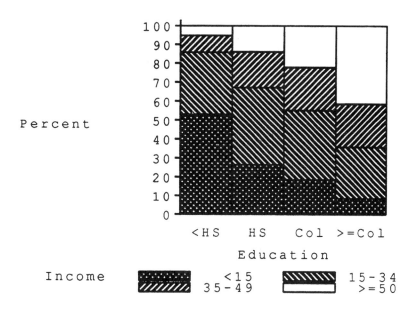

(e) Higher income is associated with more education.

2.73 48.66%, 33.73%, 10.95%, 6.66%. Approximately half of the people have one to three years of high school. About eighty percent have four years of high school or less.

2.74 Education: 25.60%, 35.60%, 16.65%, 22.15%. Income: 27.94%, 35.05%, 17.77%, 19.25%. $15-34,000.

2.75 (a) .78%, 1.65%. (b) The high blood pressure group has a higher mortality rate. The data suggests that there may be a link.

2.76 39% of the women and 64% of the men use firearms. 38% of the women and 15% of the men use poison. 13% of the women and 16% of the men use hanging. 10% of the women and 6% of the men use other means. Men are more likely than women to use firearms and women are more likely than men to use poison.

2.77 (a) (Admit, Deny). Male: (490, 210). Female: (280, 220). (b) Male: 70%, Female: 56%. (c) Business school: 80%, 90%. Law school: 10%, 33%. (d) The admission rate is the highest (83%) in the business school where the majority of the applicants (75%) are male.

2.78 (a) .263. (b) .262, .357. Yes.

2.79 More men are employed in departments where the salaries are higher.

2.80 (a) Joe: 120, 380. Moe: 130, 370. (b) .240, .260, Moe. (c)

	Right Handed Pitcher		Left Handed Pitcher	
	Hits	No Hits	Hits	No Hits
Joe	40	60	80	320
Moe	120	280	10	90

(d) Right: .400, .300, Joe. Left: .200, .100, Joe. (e) Joe hits better against both right and left-handed pitchers. Moe's batting average is higher because 80% of his at bats were against right-handed pitchers, and right-handed pitchers gave up more hits. Only 20% of Joe's at bats were against these pitchers.

2.81 (a) (Death, Not): White Defendant (19, 141); Black Defendant (17, 149). (b) Overall 12% of the whites and 10% of the blacks are sentenced to death. For white victims, 13% of whites and 17% of blacks are sentenced to death. For black victims, 0% of whites and 6% of blacks are sentenced to death. (c) The percent of death penalties for white victims is higher (14%) than for black victims (5%). Of the white defendant's victims, 94% were white.

2.82 Older children have bigger shoes and better reading comprehension. This is a common response.

2.83 Many things could cause an individual to receive more education and more income. For example, people who can easily afford more education are more likely to be in situations where there are greater opportunities for

CHAPTER 2   147

higher income. Similarly, higher innate abilities or drive to succeed, would lead to both more education and higher income.

2.84 Yes, anesthetic C may be particularly dangerous. However, anesthetic C may be used for patients who have a higher risk of death and this would confound the results.

2.85 There may be some confounding between the class of the secretaries and factors such as seniority and experience.

2.86 Not necessarily. The larger hospitals might admit the more seriously ill patients who require longer hospital stays.

2.87 Explanatory variable is serving herbal tea or not. Response variable is measure of cheerfulness. The confounding variables are time and the attention of the students.

2.88 Explanatory variable is foreign language study. Response variable is English achievement score. Students who are more skilled in languages may study foreign languages and score higher on English achievement tests.

2.89 In addition to the diet information you would want the family history and any other information that relates to blood cholesterol levels.

2.90 Type of operation and patient characteristics which may be related to risk of death.

2.91 (a) Correlation measures the strength and direction of the linear association between two quantitative variables. In this case we are quantifying the association between recalled and historical consumption. The second aim involves prediction, so regression is appropriate. **(b)** A correlation of .217 indicates a rather weak association. See plots and Figure 2.34 on page 167. **(c)** The value of r^2 is the fraction of the variation in intake at age 30 predicted by each of the explanatory variables.

2.92 (a) The overall pattern is positive linear. The point with the highest value of sodium appears to be extreme. Brands high in calories tend to be high in sodium.

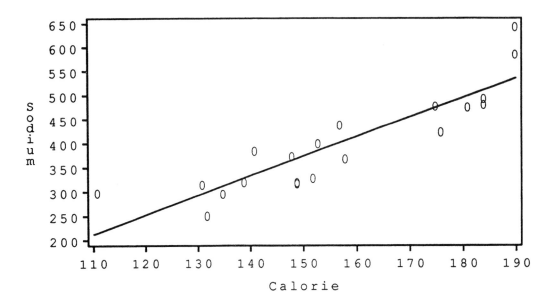

(b) $b = .8871(102.435)/22.642 = 4.01$. $a = 401.15 - (4.0133)156.85 = -228.34$. $r^2 = .7885 = 78.85\%$. **(c)** $\hat{y} = -228.34 + 4.0133(180) = 494.054$.

2.93 The plot shows no clear association. The correlation is $-.23$. There appears to be no clear relation between batting average and home runs.

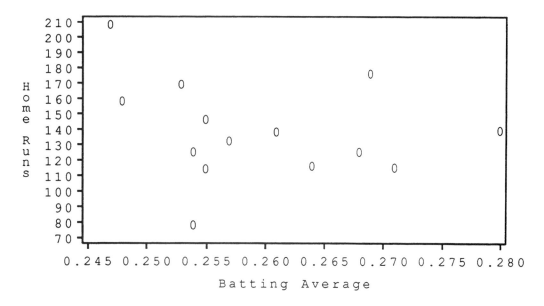

2.94 (a)

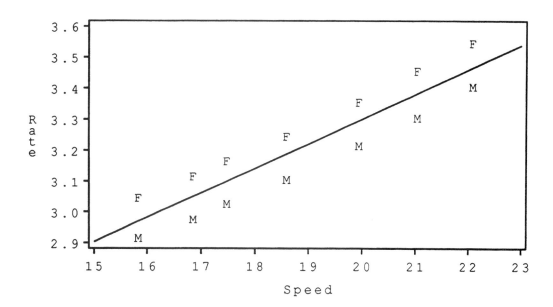

(b) $\hat{y} = 1.71 + .08x$. (c) The residuals fall into two distinct groups. The residuals are all positive for the females and all negative for the males.

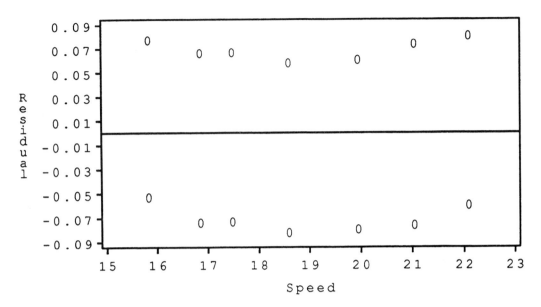

2.95 (a) Yes, the two lines appear to fit the data well. There do not appear to be any outliers or influential observations. **(b)** .189, .157. **(c)** Before: $\hat{y} = 1.089 + .189(35) = 7.704$. After: $\hat{y} = .853 + .157(35) = 6.348$. $.75(7.704 - 6.348) = \$1.017$ per day. For 31 days: $31(1.017) = \$31.53$.

2.96 (a)

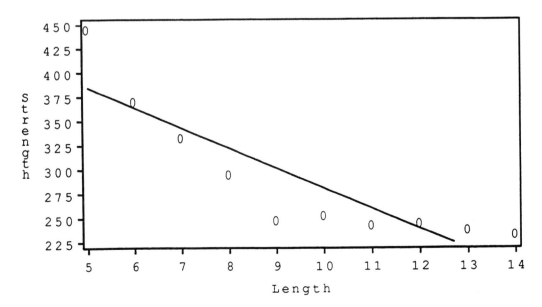

(b) As length increases strength decreases. There is a strong linear pattern from 5 to 9 inches. From 9 to 14 inches the change in strength is fairly small. There are no outliers. (c) The least squares line is $\hat{y} = 488.38 - 20.75x$. This straight line does a poor job of describing these data. One line would fit the data from 5 to 9 inches and another from 9 to 14 inches. (d) The two lines adequately describe the data. They are: 5 to 9 inches: $\hat{y} = 667.50 - 46.90x$. 9 to 14 inches: $\hat{y} = 283.10 - 3.37x$. We would ask the experts if there is some theory that would explain the leveling off of the relationship.

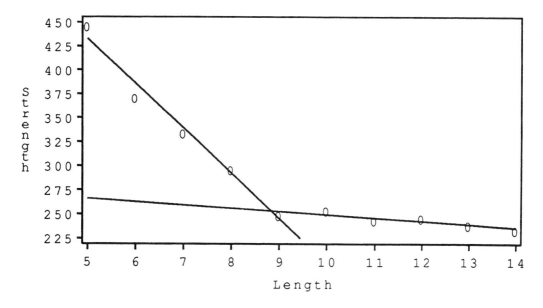

2.97 (a)

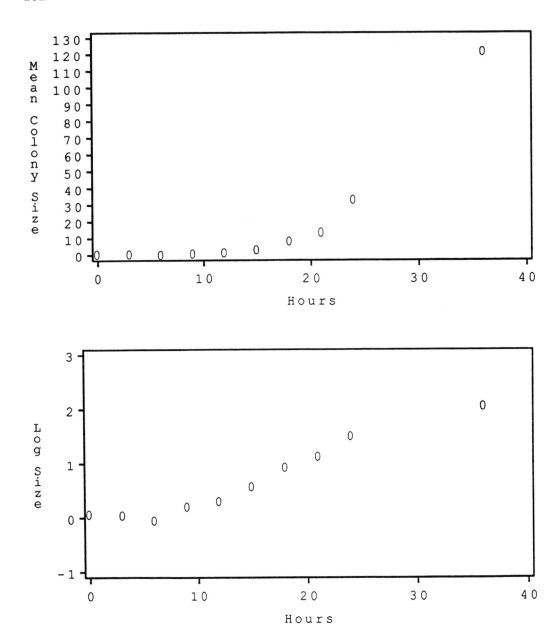

(b) Phase One (0-6 hrs) - no increase in growth. Phase Two (6-24 hrs) - exponential growth. Phase Three (36 hrs) - growth is slower than exponential. **(c)** Predicted log is .4230. Predicted colony size is 2.65.

2.98

CHAPTER 2

	Smokers			Nonsmokers	
	Overweight	Not		Overweight	Not
Early	20	60	Early	40	5
Not early	30	140	Not early	160	45

For the smokers, 40% of the overweight people die early compared to 30% of the people who are not overweight. For the nonsmokers, the percents are 20 compare to 10. For the combined data the pattern is reversed with 24% of the overweight people dying early compared to 26% of the people who are not overweight.

2.99 (a) The percentage of people voting decreases over time.

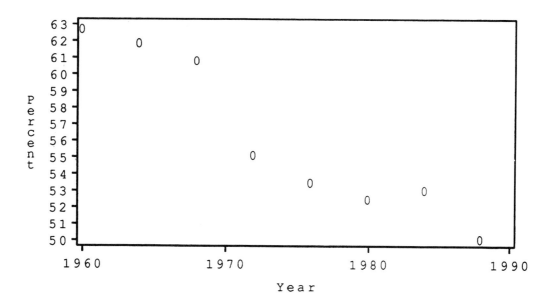

(b) The proportion of younger people eligible to vote has increased from 1960 to 1984 and younger people are less likely to vote than older people.

2.100 (a) $r = .81642$, $\hat{y} = 3.00 + .500x$ for all three sets of data. **(b)**

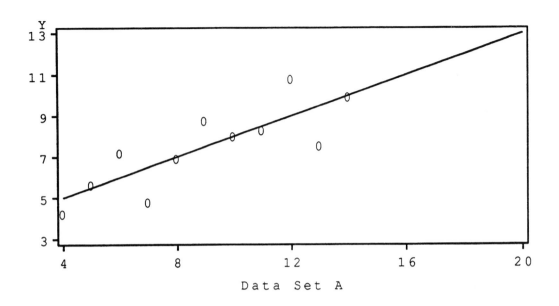

Data Set A

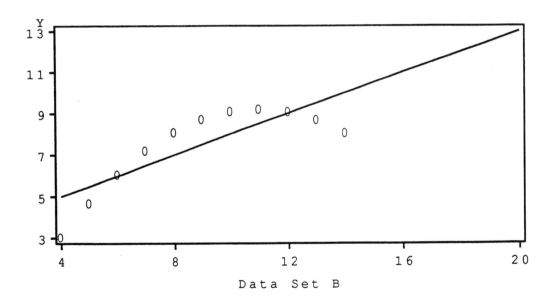

Data Set B

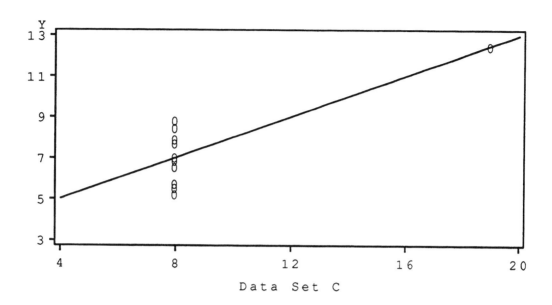

Data Set C

(c) We would be willing to use the regression line to predict y at $x = 14$ only for set A. This set shows a clear linear pattern that includes $x = 14$ as an observation. For set B, there is a curved pattern and $x = 14$ is at the end of the curve. Data set C has all of the observations at $x = 8$ with the exception of the one point at $x = 19$. There is not enough data to conclude that the pattern is linear between $x = 8$ and $x = 19$.

2.101 With case 18 omitted: $r = -.33$, $\hat{y} = 105.63 - .78x$. With case 19 omitted: $r = -.76$, $\hat{y} = 109.30 - 1.19$. With both 18 and 19 omitted: $r = -.52$, $\hat{y} = 107.59 - 1.05$. When all cases are included: $r = -.64$, $\hat{y} = 109.87 - 1.13$. Case 18 has a stronger influence on the regression line. Both cases influence the regression.

2.102 (a) The point corresponding to the veal hot dog is low on both variables but it does not deviate from the linear pattern of the other data. Therefore, we would not expect it to be very influential.

(b) Meat and beef: $\hat{y} = -162.16 + 3.62x$. Meat and beef without veal outlier: $\hat{y} = -120.65 + 3.34x$. Although the intercepts of the two lines are different, the lines are almost identical within the range of the data as shown on the plot. The case is not influential.

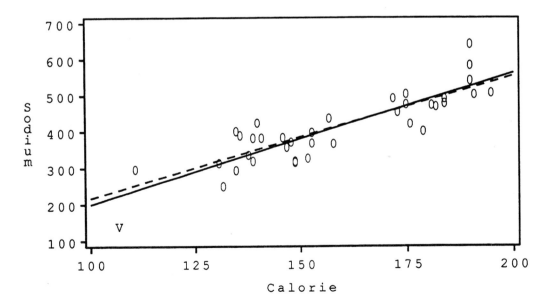

(c) The slope for poultry is less than slope for meat and beef. The intercept for poultry is greater than the intercept for meat and beef.

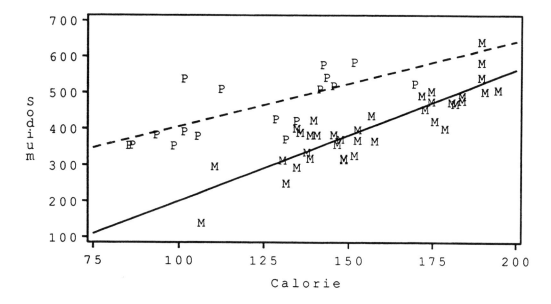

(d) The regression line for poultry is $\hat{y} = 169.23 + 2.37x$. The correlations are $r^2 = .76$, for meat and beef and $r^2 = .51$, for poultry. The squared correlation is higher for meat and beef. Therefore, we expect the prediction for this group to be more reliable. No, the regression lines are quite different.

2.103 (a) The equation is $\hat{y} = -2.58 + 1.09x$. It agrees with Exercise 2.32. (b) No clear patterns are evident. There may be one or more outliers with high positive residuals.

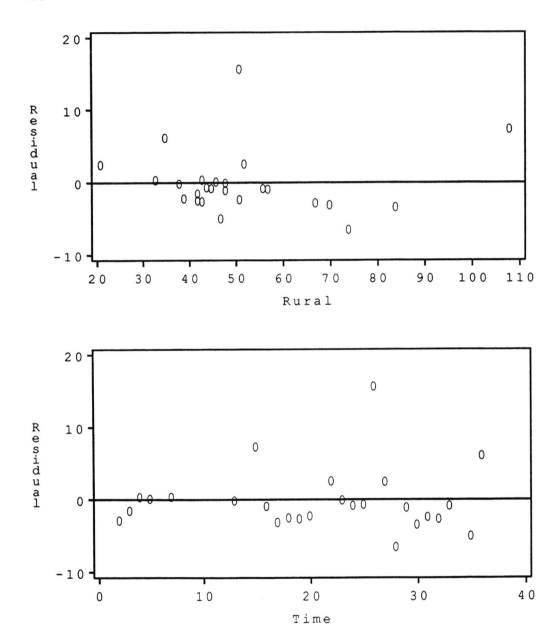

(c) This stemplot is made with the values of the residuals divided by 10 and rounded. The distribution is not symmetric; it is skewed toward high values. It is not normal.

-0 | 65

```
-0 | 33322222111111
 0 | 0000133
 0 | 67
 1 |
 1 | 6
```

2.104 (a)

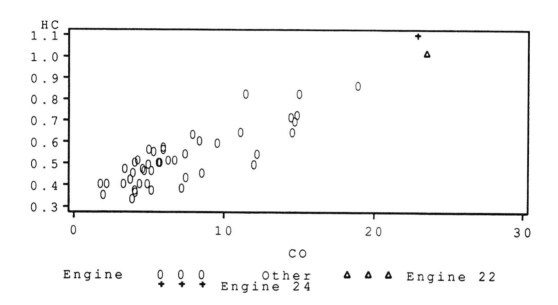

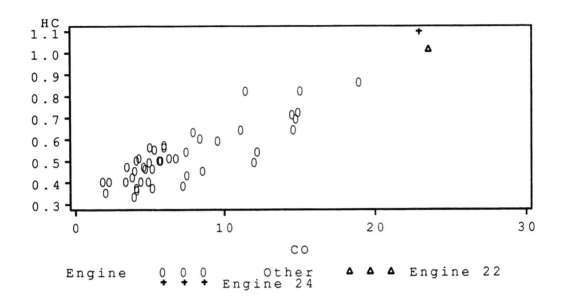

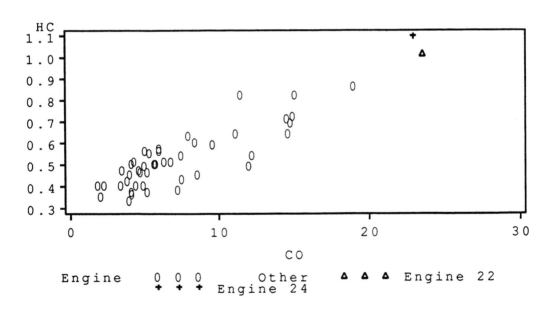

(b) HC vs CO: positive linear. HC vs NOX: negative curved. NOX vs CO: negative curved. (c) Engines 22 and 24 are unusual in all three plots. Engine 32 has a very high value of NOX. (d) All data: CO, HC: $r = .90$, CO, NOX: $r = -.69$, HC, NOX: $r = -.56$. Without 22 and 24: CO, HC: $r = .84$, CO, NOX: $r = -.69$, HC, NOX: $r = -.52$. Without 22, 24 and

32: CO, HC: $r = .84$, CO, NOX: $r = -.74$, HC, NOX: $r = -.60$. (e) All data: $\widehat{HC} = .32 + .029CO$, $\widehat{NOX} = 1.83 - .06CO$, $\widehat{HC} = .81 - .19NOX$. (f) Without 22 and 24: $\widehat{HC} = .34 + .026CO$, $\widehat{NOX} = 1.92 - .08CO$, $\widehat{HC} = .72 - .14NOX$. Without 22, 24 and 32: $\widehat{HC} = .34 + .026CO$, $\widehat{NOX} = 1.85 - .07CO$, $\widehat{HC} = .78 - .19NOX$. (g) The unusual points do not change the regression lines very much. The strongest relationship is between CO and HC.

2.105 (a) Both distributions are skewed to the right. Because the distributions are very skewed, it is difficult to distinguish between outliers and high values from a skewed distribution. Many answers are acceptable. Alaska has the highest value for pay and appears to be an outlier. DC, NY and Connecticut also have high values that are separated from the bulk of the data. There are 5 states with high values for spending (Alaska, Connecticut, DC, New Jersey and New York). New Jersey does not stand out in the pay distribution, so the same states are not outliers in both distributions. Here is the stemplot for pay:

```
21 | 3
22 | 08
23 | 0179
24 | 34
25 | 1155
26 | 379
27 | 122589
28 | 2788
29 | 04
30 | 556889
31 | 29
32 | 028
33 | 34
34 | 2
35 | 7
36 | 0146
37 |
38 | 09 (DC, NY)
39 |
```

40 | 5 (Connecticut)
41 |
42 |
43 | 2 (Alaska)

Here is the stemplot for spending:

2 | 7
3 | 02333
3 | 55678889
4 | 112222444
4 | 5666678
5 | 000111234
5 | 67789
6 | 2
6 | 5
7 | 24 (Alaska, DC)
7 | 9 (Connecticut)
8 | 14 (NY, NJ)

(b) There is a positive linear association. There is a fairly strong relationship. You would expect high spending to be associated with high salaries.

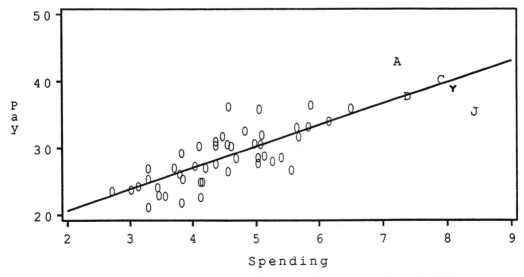

A=Alaska   D=DC   C=Connecticut   Y=NY   J=NJ

CHAPTER 2   163

(c) Many answers are acceptable. We have noted Alaska, Connecticut, DC, NJ and NY in the plot below. There are no clear outliers from the overall relationship. (d) $\hat{y} = 14.19 + 3.22x$. $r^2 = .70$. (e) $\hat{y} = 14.90 + 3.04x$. $r^2 = .69$. The lines are very similar and the values of r^2 are almost exactly the same.

2.106 (a)

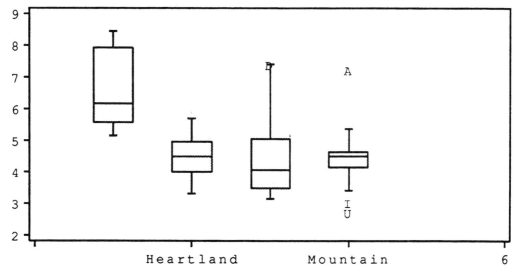

(b)

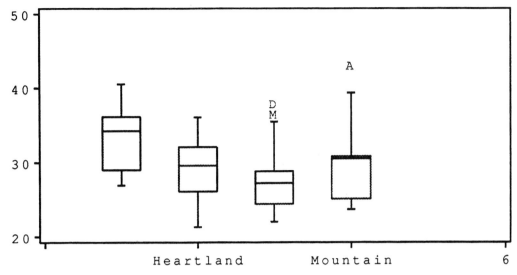

(c) There appear to be differences by region in both spending and pay. The differences are somewhat consistent for the two variables. Spending and pay tend to be higher in the Northeast than in the other regions. (d) The lease squares line is $\hat{y} = 14.19 + 3.22x$. The residuals appear to differ by region. The Northeast has low residuals and the Mountain region has high residuals. The Heartland and the South have average residuals.

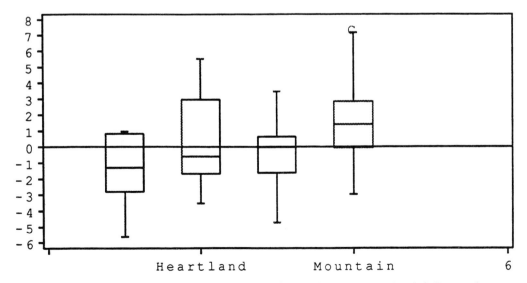

2.107 $r = .252$, $r^2 = .063$, $\hat{y} = 3.28 + .002x$. SATM is a poor predictor of GPA.

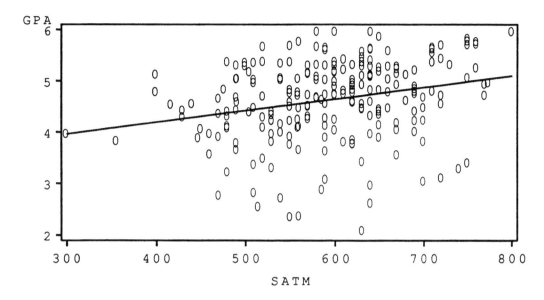

2.108 (a) There is a strong positive linear relationship. We would expect this because the same piece of wood is being measured at two different times.

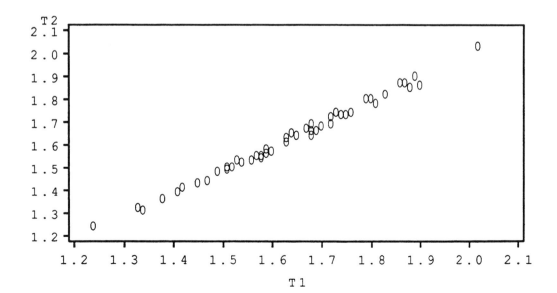

(b) $r = .996$. The measurement process is highly reliable. (c) From software, $\hat{y} = -.0333 + 1.0176$. $b = r(s_y/s_x) = .99645(.173019/.169423) = 1.0176$, $a = \bar{y} - b\bar{x} = 1.62520 - 1.0176(1.6298) = -.0333$. Note that a large number of significant digits are needed to produce accurate answers.

2.109 (a)

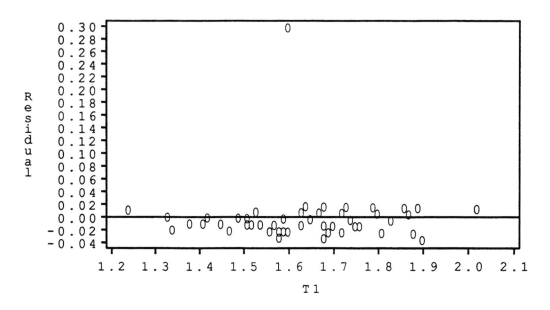

(b) Here is the stemplot for T1:

```
12 | 4
12 |
13 | 334
13 | 8
14 | 12
14 | 579
15 | 11234
15 | 678899
16 | 0003334
16 | 578889
17 | 02234
17 | 569
18 | 013
```

```
18 | 6789
19 | 0
19 |
20 | 2
```

Here is the stemplot for T2:

```
12 | 5
13 | 233
13 | 7
14 | 024
14 | 59
15 | 011344
15 | 5667889
16 | 224
16 | 5567789
17 | 00344
17 | 559
18 | 113
18 | 6788
19 | 01
19 |
20 | 4
```

(c) With the new point: $\hat{y} = -.017 + 1.011$. Without the new point: $\hat{y} = -.033 + 1.018$. The two regression lines are very similar, so the point is not influential.

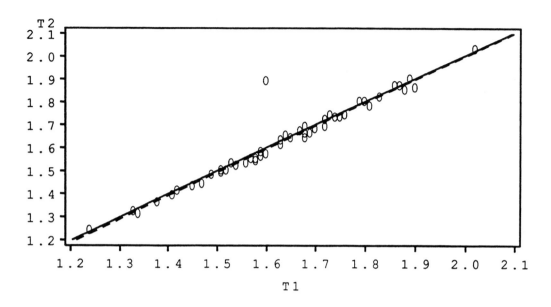

(d) With the new point the correlation is .966; without the new point it is .996.

2.110 (a) Here is the stemplot for T1:

```
12 | 4
12 |
13 | 334
13 | 8
14 | 12
14 | 579
15 | 11234
15 | 678899
16 | 003334
16 | 578889
17 | 02234
17 | 569
18 | 013
18 | 6789
19 | 0
19 |
20 | 2
20 |
```

21 |
21 |
22 | 0

Here is the stemplot for T2:

12 | 0
12 | 5
13 | 233
13 | 7
14 | 024
14 | 59
15 | 011344
15 | 5667889
16 | 224
16 | 5567789
17 | 00344
17 | 559
18 | 113
18 | 6788
19 | 1
19 |
20 | 4

(b) T1 With: $\bar{x} = 1.641$, $s = .186$. T1 Without: $\bar{x} = 1.630$, $s = .169$. (b) T2 With: $\bar{x} = 1.617$, $s = .181$. T2 Without: $\bar{x} = 1.625$, $s = .173$. The new point has very little effect on these summary measures. (c) With: $\hat{y} = .482 + .692x$. Without: $\hat{y} = -.0333 + 1.018x$. Yes the new point is influential; it changes the least squares line substantially.

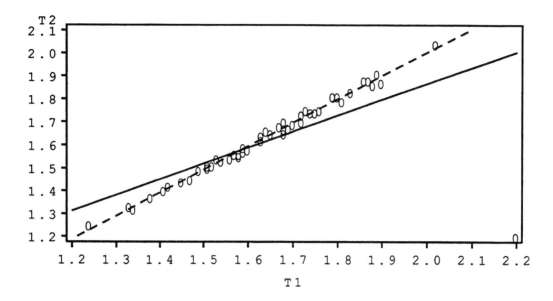

(d) With: $r = .709$. Without: $r = .996$. The new point changes the correlation substantially.

4.3 CHAPTER 3

3.1 This is not an experiment. The okra may have been planted in an area of the garden which is not attractive to stinkbugs.

3.2 This anecdotal evidence is based on a sample of size one and is not an experiment.

3.3 (a) Treatments are not actively imposed. (b) This is a survey and no treatment is imposed.

3.4 (a) The explanatory variable is the type of mastectomy and the response variable is survival time. (b) The explanatory variable is sex and the response variable is whether or not they voted for the Democratic or Republican candidate.

3.5 No. No treatment is imposed. The explanatory variable is whether or not they live in public housing. The response variable is family stability and other variables.

3.6 No it is not an experiment because no treatment is imposed.

3.7 Yes. The treatment is walking briskly on the treadmill. The fact that eating is not recorded limits the conclusions that can be drawn. The explanatory variable is time after exercise, and the response variable is the metabolic rate.

3.8 Yes this is an experiment. The treatment is that the subjects are asked to taste the two brands of muffins.

3.9 Experimental units: pairs of pieces of package liner. Explanatory variable: temperature of jaws. Response variable: peel strength of the seal.

3.10 Experimental units: male physicians. Explanatory variable: aspirin or placebo. Response variable: heart attack or not.

3.11 Experimental units: one day old male chicks. Explanatory variables:

corn variety and protein level. Response variable: weight gain.

3.12 The treatments are aspirin every second day and placebo. The subjects are randomly assigned to two groups; the treatment is then administered to each group and the observation is taken several years later.

3.13 250°, 275°, 300°, 325°.

3.14 There are nine treatments: opaque-2 with 12%, 16% and 20% protein, floury-2 with 12%, 16% and 20% protein, and normal with 12%, 16% and 20% protein. Randomly assign ten chicks to each of nine groups; the diet is then fed to the chicks and the weight gain is measured 21 days later.

3.15 (a) Patients could be randomized to the two treatments. (b) No, there is no treatment that can be applied.

3.16 (a) Randomly assign 20 subjects to each of the two treatment groups; administer the treatment and after one hour ask the subjects to estimate the percentage of pain relieved. (b) Assign the number 00 to Abrams, 01 to Adamson, etc. The first twenty subjects selected will receive the medication treatment. These were selected in the following order: Chen, Abrams, Rosen, Iselin, Cansico, Martinez, Obramowitz, Janle, Gutierrez, Ullmann, Wong, Roberts, Hwang, Curzakis, Gerson, Solomon, Afifi, Turing, Lattimore, Morse. The remaining subjects will receive the placebo treatment.

3.17 Assume the field has four rows and five plots per row. First ten numbers between 1 and 20 correspond to plots receiving treatment A. Row 1: AABBB, Row 2: AAAAA, Row 3: BBBBB, Row 4: ABBAA.

3.18 Number the boards from 01 to 16. The first four numbers selected are assigned to lemon yellow, the next four to white, etc. The field with the pole number and color looks as follows:

1B	2G	3L	4L
5B	6W	7G	8B
9L	10W	11W	12B
13G	14L	15G	16W

3.19 250°: Units 4, 7, 10, 16, 19; 275°: Units 5, 8, 9, 13, 15; 300°: Units 1, 3, 6, 11, 18; 325°: Units 2, 12, 14, 17, 20.

3.20 Randomly assign two classes to each of the six treatments; students are taught using the instructional unit and a final test is given. The classes are numbered from 00 to 11. The assignment of classes to treatments is as follows:

	Fact	Computation	Word Problem
Before	05, 00	01, 04	09, 07
After	02, 08	11, 06	03, 10

3.21 First two numbers between 1 and 20 correspond to plots receiving treatment A and genetic line one, next two numbers correspond to plots receiving treatment A and genetic line two, etc. Row 1: A3, A5, B2, B4, B3; Row 2: A1, A5, A4, A2, A2; Row 3: B3, B2, B4, B5, B1; Row 4: A3, B5, B1, A1, A4.

3.22 By randomizing, such factors as age and medical history are not confounded with the results.

3.23 The experimenter was aware of which subjects were assigned to the meditation group, and might rate these subjects as having lower anxiety.

3.24 The results cannot be trusted because the results are confounded with the person who ran the assays.

3.25 For each person, flip the coin. If heads, measure the right hand and then the left hand. If tails, measure in reverse order.

3.26 (a) and (b) Within each block assign the numbers 1 to 4 to the four subjects. The first number selected will receive diet A, the second diet B, etc. Random numbers for block one were obtained from line 101 of Table B, block two from line 102, etc. Block 1: 22-Williams-A, 24-Festinger-B, 25-Hernandez-C, 25-Moses-D; Block 2: 27-Siegel-C, 28-Kendall-D, 28-Mann-A, 29-Smith-B; Block 3: 30-Brunk-B, 30-Orbach-D, 30-Rakow-C, 32-Loren-A; Block 4: 33-Jackson-B, 33-Stall-A, 34-Brown-C, 34-Dixon-

D; Block 5: 35–Birnbaum–C, 35–Suggs–A, 39–Nevesky–D, 42–Wilansky–B.

3.27 First, assign the numbers 1 to 6 to the schools in district one. From line 125 of Table B we use the following digits for the randomization: 6, 4, 1, 2, 3, 5. Assign school number 6 to Fact-Before, 4 to Compute-Before, 1 to Word-Before, 2 to Fact-After, 3 to Compute-After, and 5 to Word-After. For district two the randomization is 1, 6, 2, 3, 4, 5.

3.28 In each field there will be two poles of each color. For field one label the poles from 1 to 8. The first two numbers selected will receive lemon yellow, the next two white, etc. Do the same for field two starting at line 106. Note there are many correct ways to do this exercise that will give different results.

```
Field One   1G    2L    3B    4W
            5L    6G    7W    8B
Field Two   1W    2B    3G    4W
            5B    6L    7G    8L
```

3.29 **(a)** The basic unit is a person. The population is all adult U.S. residents. We assume that an adult is any person 18 years of age or older. **(b)** The basic unit is a household. The population is all U.S. households. **(c)** The basic unit is a voltage regulator. The population is all voltage regulators from the last shipment.

3.30 **(a)** The population is American male college students enrolled in Sociology 101. **(b)** The Congressman would like the population to be all of his constituents. However, the people who wrote to the Congressman are voluntary respondents who do not represent this population.
(c) The population is all auto insurance claims in a particular month.

3.31 The sample was not representative of women in general because the questionnaires were distributed through women's groups and the response was voluntary. It is known that voluntary respondents are more likely to have stronger feelings about their opinions than nonrespondents. It is likely that these facts would cause the percents to be higher.

3.32 The sample is biased because of voluntary response. The percent in the population who oppose gun control is probably higher.

3.33 Beginning with A1096 and going across rows, label the control numbers with the numbers from 1 to 25. From line 111 of Table B we select the following: 12-B0986, 04-A1101, 11-A2220.

3.34 Selecting three digit numbers from Table B gives the following sample: 214, 313, 409, 306, 511.

3.35 Beginning with Agarwal and going down the columns, label the people with the numbers 1 to 28. From line 139 of Table B we select the following: 04-Bowman, 10-Frank, 17-Mihalko, 19-Naber, 12-Goel, 13-Gupta.

3.36 If we begin at the same place, we get the same randomization every time. For example, in Exercise 4.28 we want a different randomization for each field.

3.37 (a) We will choose one of the first 40 at random and then the addresses 40, 80, 120, and 160 places down the list from it. The addresses selected are 35, 75, 115, 155, 195. (b) Because we choose 1 of the first 25 at random, each of the first 25 has the same chance to be chosen. But each of the second 25 is chosen exactly when the corresponding address in the first 25 is chosen, and so on. So each address has the same chance of begin chosen. This is not an SRS because the only possible samples have exactly 1 address from the first 25, 1 address from the second 25, and so on. An SRS could contain any 5 of the 200 addresses in the population. Note that his view of systematic sampling assumes that the number in the population is a multiple of the sample size.

3.38 Assign the numbers 1 to 30 to the students. The first four numbers selected are 04, 17, 22, 09 starting at line 115 of Table B. These correspond to David, Klotz, O'Brien, Griswold. Assign the numbers 1 to 10 to the faculty. The first two numbers selected are 04 and 07 starting at line 120. These correspond to Gupta and Moore.

3.39 Give each name on the alphabetized lists a number; 001 to 500 for females and 0001 to 2000 for males. From line 122 of Table B, the first five females selected are 138, 159, 052, 087, and 359. If we start at the beginning of line 122 of Table B, the first five males selected are 1387, 0529, 0908, 1369, 0815.

3.40 No, in SRS all samples have the same chance of being selected. This is a stratified sample. An SRS would be very unlikely to select exactly 200 males and 50 females.

3.41 (a) Households without telephones and those with unlisted numbers are omitted. These households would include people who cannot afford a telephone, people who choose not to have a telephone, and people who prefer not to list their telephone number. (b) The random digit method includes unlisted numbers in the sampling frame.

3.42 (a) Only households with someone at home during the day are sampled. We would expect this subset of households to have a higher percentage of people who bake their own bread. (b) The presence of a uniformed police officer would probably cause people to respond differently.

3.43 (a) Many people think that "food stamps" refer to the prize stamps that some grocery stores give away. So the question is not clear. (b) The question is clear but it is slanted because it asks to choose between an extreme position (confiscate) and a general statement taken from the Constitution. (c) The question is clear but it is slanted toward agreement because it gives reasons to support a freeze. (d) The question is unclear. It uses fancy words and technical phrases that many people will not understand. It is slanted toward a positive response because it gives reasons to favor recycling.

3.43 Statistic.

3.45 Parameter, statistic.

3.46 Statistic, parameter.

3.47 Statistic, statistic.

3.48 (a) When we tossed the coin 20 times we observed 8 heads. The value of $\hat{p}$ is $8/20 = .40$. (b) Here is a stem plot of the 10 values of $\hat{p}$ that we observed.

```
2 | 5
3 |
3 |
4 | 00
4 |
5 | 00
5 | 55
6 | 000
```

3.49 (a) Using line 120 of Table B, we select students 3, 5, 4 and 7. Their scores are 58, 73, 72 and 66. The mean is 67.25. **(b)** The means are 67.25, 66.50, 68.00, 70.00, 75.25, 69.00, 67.25, 69.00, 64.50 and 71.00. The mean of our ten repetitions (for each repetition we select 4 students and calculate the mean score) is 68.775. This mean is quite close to the population mean 69.4. Here is a stemplot of our results.

```
64 | 5
65 |
66 | 5
67 | 22
68 | 0
69 | 00
70 | 0
71 | 0
75 | 2
```

3.50 We let the digits 0 and 1 represent the presence of eggs and the digits 2 to 9 represent the absence of eggs. We use ten digits for each sample. Starting at line 135, the first 10 digits are 66925 55658. Since there are no 0s or 1s, there are no egg masses in the first 10 sample areas. The value of $\hat{p}$ is 0. **(b)** We continue the process using lines 136 to 154. Note that lines 101 to 150 of Table B can ve found on the inside of the back cover; but for lines 151 to 200, you need to look in the tables section of the text. The sample proportions are 0, .2, .1, .2, .1, .2, 0, .1, 0, .1, .2, .2, 0, .3, .2, .4, .3, .3, .1, and .2. The mean is .16. This distribution shows some skewness toward high values. For the stemplot we view the observations as having two digits after the decimal (e.g. .2 is .20). We use the first digit for the stem and the second for the leaf. Note results will vary depending on what part of Table B you use.

```
0 | 0000
1 | 00000
2 | 0000000
3 | 000
4 | 0
```

3.51 (a) Use digits 0, 1, 2, 3 for "Yes" and 4 to 9 for "No". Starting at the beginning of line 110, we obtain $\hat{p} = .25$. **(b)** For lines 111 to 119, the values of $\hat{p}$ are .30, .30, .45, .40, .40, .45, .50, .50, .25. The mean is .38. Yes.

3.52 (a) (freq, $\hat{p}$): (1, .045), (3, .065), (2, .070), (5, .075), (11, .080), (12, .085), (12, .090), (9, .095), (7, .100), (5, .105), (6, .110), (7, .115), (10, .120), (4, .125), (1, .130), (2, .135), (2, .140), (1, .150). **(b)** Median=.095, Mean=.098. It is approximately normal. The statistic $\hat{p}$ is an unbiased estimator of p. **(c)** The center of the sampling distribution would be at .10. The spread would be smaller because the sample size is larger.

3.53 (a) Large bias, large variability. **(b)** Small bias, small variability. **(c)** Small bias, large variability. **(d)** Large bias, small variability.

3.54 Larger samples produce statistics with less variability.

3.55 For this exercise we assume the population proportions in all states are about the same. The effect of the population proportion on the variability will be studied further in Chapter 6. **(a)** No. The variability is controlled by the size of the sample. **(b)** Yes. The sample sizes will vary from 2100 in Wyoming to 120,000 in California and larger samples are less variable.

3.56 The margin of error would be the same in both cases. As long as the population is much larger than the sample, the spread of the sampling distribution is controlled entirely by the size of the sample. The answers assume that the population proportions are the same in both cases.

CHAPTER 3

3.57 For each taster flip a coin. If heads taste Pepsi first, then Coke. If tails taste Coke first, then Pepsi.

3.58 The letters will be mailed at 9 am, noon and 5 pm. There are six treatment combinations. For each letter randomize the day of the week on which it will be sent.

3.59 You want to compare how long it takes to walk to class by two different routes. The experiment will take 20 days. The days are labeled from 01 to 20. Using Table B, the first 10 numbers between 01 and 20 will be assigned to route A; the others will be assigned to route B. Take the designated route on each day and record the time to get to class. Note that this experiment is not blind; you know the route that you take on each day.

3.60 (a) Randomize to determine the order the tests will be taken. **(b)** Use 22 digits from Table B. If the digit is between 0 and 4 administer BI and then ARMSA; otherwise use the reverse order.

3.61 (a) Fifteen patients are randomly assigned to receive β-blockers. The other 15 patients will receive a placebo. **(b)** The patients are numbered from 01 to 30. Those receiving the β-blockers are 21, 18, 23, 19, 10, 08, 03, 25, 06, 11, 15, 13, 24, 09 and 28.

3.62 No, it is an observational study. No treatment is imposed.

3.63 (a) Assign the numbers 0001 to 3478 to the alphabetized list of students. **(b)** 2940, 0769, 1481, 2975, 1315.

3.64 Use a stratified sample with 50 faculty selected from each class. This plan neglects the particular college where an individual faculty member is employed. An alternate scheme would be to sample colleges, and then faculty within colleges.

3.65 (a) The population is all students classified as full-time undergraduates at the beginning of the Fall semester on a list provided by the Registrar. **(b)** Stratify on class and randomly select 125 students in each class. **(c)** If questionnaires are mailed some students will not respond. If telephones are used, the students without a telephone will not be contacted.

3.66 (a) Randomly assign black men to the two treatments. Do the same for white men. Measure blood pressure when the subjects enter the study. Repeat the measurement after the treatment period is complete and determine the change in blood pressure. **(b)** With larger sample sizes the variability in our estimates is decreased.

3.67 (a) There are two factors and six treatments. The treatments are 50°-60 rpm, 50°-90 rpm, 50°-120 rpm, 60°-60 rpm, 60°-90 rpm, 60°-120 rpm. 12 experimental units. **(c)** Label the batches with the numbers 01 to 12. The first two numbers that appear in the table will be given treatment 50°-60 rpm, the next two will be given 50°-90 rpm, etc. The assignment is 06 and 09 to 50°-60 rpm, 03 and 05 to 50°-90 rpm, 04 and 07 to 50°-120 rpm, 02 and 08 to 60°-60 rpm, 10 and 11 to 60°-90 rpm, 12 and 01 to 60°-120 rpm.

3.68 (a) Assign the labels 01 to 20 to the twenty subjects. Using pairs of digits from Table B, let the first 10 labels that appear be assigned to the 70 deg condition. The other 10 subjects are assigned to the 90 deg condition. The subjects perform the task under the assigned temperature conditions and the number of correct insertions is recorded. **(b)** In the matched pairs experiment each subject performs the task under both temperature conditions. The order in which the two conditions are used should be randomized. Use Table B or flip a coin. Another alternative would be to randomly assign the subjects to two groups. The first group would do the 70 deg first and then the 90 deg condition. The other group would use the reverse order. In this experiment it is very important that the subjects are fully trained before collecting the data. If they are not, then learning effects may be a problem.

3.69 If the patients who choose not to participate in this study are in some way different from those who do choose to participate, then including them in the control group would bias the estimation for that group.

3.70 The results have no statistical validity because the voluntary letter writers are not necessarily representative of voters in general.

3.71 This is an observational study and no treatment is imposed. There is no information to support the claim that you can alter your moods if you start to run.

3.72 In Example 3.8, 4 rats with genetic defects are in the experimental group and 6 are in the control group.

3.73 In Minitab use the code given in the problem with 30 replaced by 150 in both places. The subjects with the first 25 labels in the randomly ordered list are assigned to treatment 1. The next 25 labels correspond to treatment 2. Continue to the last 25 which are assigned to treatment 6. In most software packages you can use a version of the following method. Generate cases with the numbers 1 to 150. Generate a random number for each case (usually this will be a uniform random variable on zero to one; Example 4.19 on page 312 discusses this distribution). Then sort the cases on the random number and divide the groups as described above.

3.74 Both the center and the spread change with p. The variability increases as p goes from .1 to .3 to .5. The distributions are approximately normal.

3.75 Increasing the size of the sample decreases the variability of the statistic. The mean remains the same.

4.4 CHAPTER 4

4.3 (a) $S = \{germinates, \ does \ not \ germinate\}$. **(b)** If measured in weeks, for example, $S = \{0, 1, 2, \ldots\}$. **(c)** $S = \{A, B, C, D, F\}$. **(d)** $S = \{makes \ the \ shot, \ misses \ the \ shot\}$. **(e)** $S = \{1, 2, 3, 4, 5, 6, 7\}$.

4.4
(a) $S = \{all \ numbers \ between \ 0 \ and \ 24\}$. **(b)** $S = \{0, 1, \ldots, 11,000\}$. **(c)** $S = \{0, 1, \ldots, 12\}$. **(d)** $S = \{all \ numbers \ greater \ than \ or \ equal \ to \ zero\}$. **(e)** $S = \{all \ positive \ and \ negative \ numbers\}$. Note the rats can lose weight.

4.5 $S = \{all \ numbers \ between \ 0.00 \ and \ 5.00\}$.

4.6 $S = \{all \ numbers \ between \ 0 \ and \ 300\}$.

4.7 $1 - (.49 + .27 + .20) = .04$.

4.8 $1 - (.3 + .2 + .2 + .2 + .1) = 0$.

4.9 Model 1: Legitimate. Model 2: Legitimate. Model 3: Sum is less than one. Model 4: Probabilities cannot be negative.

4.10 (a) Legitimate. **(b)** Not legitimate because the sum is greater than one. **(c)** Not legitimate because the sum is less than one.

4.11 No. P(NCS) = P(V) = 1. P(D) = 3(.1) = .3. P(NC) = 2P(D) = .6. The sum of the probabilities is greater than one.

4.12 (a) $.3 + .2 = .5$. **(b)** $.1 + .2 + .1 = .4$. **(c)** $1 - .2 = .8$. **(d)** $1 - (.1 + .1) = .8$. **(e)** 1.

4.13 $1 - .46 = .54$.

4.14 $1 - .36 = .64$.

4.15 $.45 + .22 = .67. 1 - .67 = .33$.

CHAPTER 4

4.16 $.20 + .28 = .48$. $1 - .48 = .52$.

4.17 (a) $.41 + .23 = .64$. $.06 + .01 = .07$. **(b)** The event that the student selected ranked in the lower 60% in high school. $.29 + .06 + .01 = .36$. $1 - .64 = .36$. **(c)** $.41 + .23 + .06 + .01 = .71$. $.64 + .07 = .71$.

4.18 (a) $.08 + .20 = .28$. $.09 + .05 + .03 = .17$. **(b)** The farm is 50 acres or more in size. $1-.28=.72$. **(c)** The farm is less than 50 acres or 500 acres or more in size. $.28+.17=.45$.

4.19 (a) $.14+.11+.05+.11+.12+.025+.11+.20+.08+.01+.04+.005 = 1$. **(b)** $.11 + .20 + .08 + .01 + .04 + .005 = .445$. **(c)** $1 - (.025 + .005) = .97$. **(d)** $.11 + .01 + .12 + .04 = .28$. **(e)** $1 - .28 = .72$.

4.20 (a) 1/38. **(b)** 18/38. **(c)** 12/38.

4.21 (a) S = {(Abby, Deborah), (Abby, Jim), (Abby, Julie), (Abby, Sam), (Deborah, Jim), (Deborah, Julie), (Deborah, Sam), (Jim, Julie), (Jim, Sam), (Julie, Sam)} **(b)** There are 10 equally–likely choices, so the probability is .1 **(c)** Four of the choices include Julie, so the probability is .4. **(d)** This corresponds to the choices (Abby, Deborah), (Abby, Julie), and (Deborah, Julie). The probability is .3.

4.22 The probability that all three small battles are won is $.8^3 = .51$. This is less than the probability of winning one large battle (.6), so the strategy is to fight one large battle.

4.23 $.95^{12} = .540$.

4.24 No. This calculation assumes that the events are independent. We would expect laborers and operators to be less likely that the general population to have at least 4 years of college.

4.25 $(1 - .02)^{20} = .668$.

4.26 Consider the first five rolls. Each of the three sequences has one G and four R's so the probabilities are the same. For the first sequence you win if the sixth roll is either R or G; for the second you win only if it is G; and for the third you win only if it is R. Therefore, the first is more likely. The three probabilities are $(2/6)^5(4/6) = .0082$, $(2/6)^5(4/6)^2 = .0055$, $(2/6)^6(4/6) = .0027$.

4.27 The father has one A gene and one B gene. Each of these has probability of .5 of being passed to the child. The same is true for the mother. The probability that the child receives two A genes is $(.5)(.5) = .25$. Similarly the probability that the child receives two B genes is $(.5)(.5) = .25$. The child will have blood type AB if the mother contributes an A gene and the father contributes a B gene or vice versa. Each of these events has probability $(.5)(.5) = .25$. Therefore, the probability that the child has blood type AB is $.25 + .25 = .50$.

4.28 (a) $.65^3 = .2746$. (b) The events are independent so the probability is $1 - .65 = .35$. (c) The probability that it moves up in both of the next two years is $.65^2 = .4225$. The probability that it moves down in both of the next two years is $.35^2 = .1225$. Therefore, the probability is $.4225+.1225=.5450$.

4.29 The probabilities in the table are obtained by multiplying the marginal probabilities. For example, the first entry is $.090(.614) = .055$. Dem: .055, .293, .265. Rep: .035, .185, .167. Note that the sums of the columns differ from the given marginal sums by .001. The difference is due to roundoff.

4.30 (a) $.321+.124 = .445$; $.365+.190$ (or $1-.445$) $= .555$. (b) $.321+.365 = .686$; $.124+.190$ (or $1-.686$) $= .314$. (c) If the events are independent that the probability that the patient is 65 or older and tests were done is $(.555)(.686) = .381$. The true probability is .365, so the events are not independent. Since the true probability is less than that calculated assuming independence, we conclude that older patients are tested less frequently than they would be if testing were independent of age.

4.31 (a) All probabilities are between 0 and 1. The sum of the probabilities is $.48+.38+.08+.05+.01 = 1.00$.

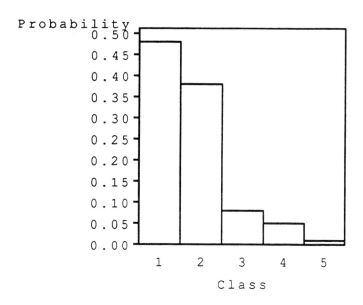

(b) .48+.38+.08 = .94. **(c)** .48+.38 = .86. **(d)** $X \geq 4$. .05 + .01 = .06. **(e)** $X > 1$. 1−.48 = .52.

4.32 (a) All probabilities are between 0 and 1. The sum of the probabilities is .240+.322+.177+.155+.067+.024+.015 = 1.00.

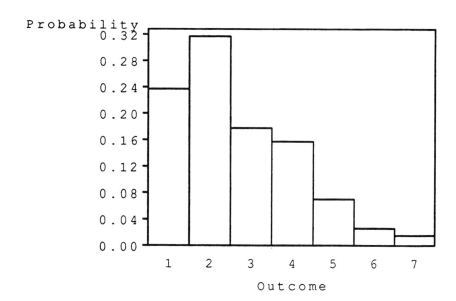

(b) .067+.024+.015 = .106. (c) .024+.015 = .039. (d) .177+.155 = .332.
(e) 1−.240 = .760. (f) $X > 2$. .177+.155+.067+.024+.015 = .438.

4.33 (a) All probabilities are between 0 and 1. The sum of the probabilities is .010+.007+.007+.013+.032+.068+.070+.041+.752 = 1.00.

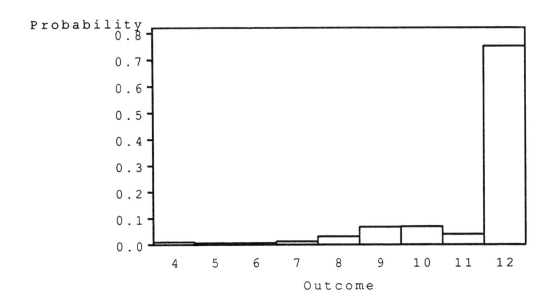

(b) 1−(.010+.007) = .983. (c) 1−(.010+.007+.007) = .976. (d) $X \geq 9$. .068+.070+.041+.752 = .931. (e) $X < 12$. 1−.752 = .248.

4.34 (a) $X = 0$: BBBB. $X = 1$: GBBB, BGBB, BBGB, BBBG. $X = 2$: GGBB, GBGB, GBBG, BGGB, BGBG, BBGG. $X = 3$: GGGB, GGBG, GBGG, BGGG. $X = 4$: GGGG. **(b)** 15/16 = .9375. **(c)** 6/16 = .3750. **(d)** 8/16 = .50.

4.35 (a) (1,1), (1,2), (1,3), (1,4), (1,5), (1,6), (2.1), (2,2), (2,3), (2,4), (2,5), (2,6), (3,1), (3,2), (3,3), (3,4), (3,5), (3,6), (4,1), (4,2), (4,3), (4,4), (4,5), (4,6), (5,1), (5,2), (5,3), (5,4), (5,5), (5,6), (6,1), (6,2), (6,3), (6,4), (6,5), (6,6). **(b)** 1/36. **(c)** (x, prob): (2, 1/36), (3, 2/36), (4, 3/36), (5, 4/36), (6, 5/36), (7, 6/36), (8, 5/36), (9, 4/36), (10, 3/36), (11, 2/36), (12, 1/36).

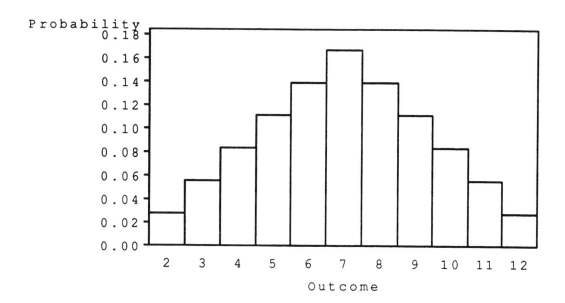

(d) $6/36+2/36 = 2/9$. (e) $1-6/36 = 5/6$.

4.36 (a) $.6(.6)(.4) = .144$. (b) Possible combinations are: SSS, SSO, SOS, SOO, OSS, OSO, OOS, OOO. The probabilities are $.6^3 = .216$, $.6^2(.4) = .144$, $.144$, $.6(.4^2) = .096$, $.144$, $.096$, $.096$, $.4^3 = .064$.
(c)

x_i	0	1	2	3
p_i	.216	.432	.288	.064

The probabilities are calculated by summing the probabilities from part (b). For example, $P(X = 0) = P(SSS) = .216$; $P(X = 1) = P(SSO, SOS, OSS) = 3(.144) = .432$. (d) $X \geq 2$. $.288+.064 = .352$.

4.37 (a) $.4-0 = .4$. (b) $1-.4 = .6$. (c) $.5-.3 = .2$. (d) $.5-.3 = .2$. (e) $.713-.226 = .487$.

4.38 (a) $.49-0 = .49$. (b) $1-.27 = .73$. (c) $1-.27 = .73$. Note the probability that $X > 1$ is zero. (d) $(.2-.1)+(.9-.8) = .2$. (e) $(.3-0)+(1-.8) = .5$. (f) 0.

4.39 (a) Since the area under the curve must be 1, two times the height must be 1. Therefore, the height is .5.

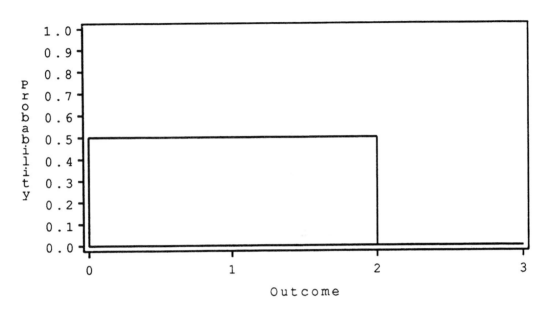

(b) $(1-0)(.5) = .5$. (c) $(1.3-.5)(.5) = .4$. (d) $(2.0-.8)(.5) = .6$. (e) $(2.0-1.4)(.5) = .3$. Note $P(X > 2) = 0$. (f) 0.

4.40 (a) .8413. (b) $1-.1587 = .8413$. (c) 0.

4.41 (a) $.8413-.1587 = .6826$. (b) $.9772-.8413 = .1359$. (c) $.9772-.8413 = .1359$.

4.42 (a) $P(\hat{p} \geq .5) = P(Z \geq (.5 - .3)/.023) = P(Z \geq 8.70) = 0$. (b) $P(\hat{p} < .25) = P(Z < (.25 - .3)/.023) = P(Z < -2.17) = .0150$. (c) $P(.25 < \hat{p} < .35) = P((.25 - .3)/.023 < Z < (.35 - .3)/.023) = P(-2.17 < Z < 2.17) = .985 - .015 = .970$. (d) $P(\hat{p} \leq .4 \text{ or } \hat{p} \geq .6) = P(Z \leq (.4 - .3)/.023) + P(Z \geq (.6 - .3)/.023) = P(Z \leq 4.35) + P(Z \geq 13.04) = 1.00 + 0 = 1.00$.

4.43 (a) $P(\hat{p} = .16) = 0$. (b) $P(\hat{p} \geq .16) = P(Z \geq (.16 - .15)/.0092) = P(Z \geq 1.09) = 1 - P(Z \leq 1.09) = 1 - .8621 = .1379$. (c) $P(.14 \leq \hat{p} \leq .16) = P(\hat{p} \leq .16) - P(\hat{p} \leq .14) = P(Z \leq (.16 - .15)/.0092) - P(Z \leq (.14 - .15)/.0092) = P(Z \leq 1.09) - P(Z \leq -1.09) = .8621 - .1379 = .7242$.

4.44 (a) There may be some confusion regarding terms such as "payoff" and

CHAPTER 4

"win". The player gives the house $1. If the player's number is drawn, the house gives the player $3. This is the payoff. If the player's number is not drawn the house gives the player nothing; so the payoff is 0.

Payoff	$0	$3
Probability	.75	.25

(b) The mean payoff is $\mu_X = (0)(.75) + (3)(.25) = .75$. **(c)** Since the price of the bet is $1, the casino keeps $1 minus .75 or 25 cents.

4.45 $1(.240) + 2(.322) + 3(.177) + 4(.155) + 5(.067) + 6(.024) + 7(.015) = 2.619$.

4.46 $4(.010) + 5(.007) + 6(.007) + 7(.013) + 8(.032) + 9(.068) + 10(.070) + 11(.041) + 12(.752) = 11.251$.

4.47 $0(.10) + 1(.15) + 2(.30) + 3(.30) + 4(.15) = 2.25$.

4.48 $1 - (.00183 + .00186 + .00189 + .00191 + .00193) = .99058$.
$\mu_x = (-99,750)(.00183) + (-99,500)(.00186) + (-99,250)(.00189) + (-99,000)(.00191) + (-98,750)(.00193) + (1250)(.99058) = 303.36$.

4.49 The company will earn $309.74 per person in the long run because of the law of large numbers.

4.50 There are 1000 different three digit numbers (000 to 999); so the probability of each is $1/1000 = .001$. There are six arrangements of three unique digits. You win if any of the six arrangements is chosen. Therefore, the probability of winning is .006 nd the probability of losing is $1 - .006 = .994$. The expected payoff is $(83.33)(.006) + (0)(.994) = .50$.

4.51 (a) Independent. Weather conditions a year apart should be independent. **(b)** Not independent. Weather patterns persist for several days so we would expect some dependence here. **(c)** Not independent. The two locations are close together so the weather should be similar in both locations.

4.52 (a) There are two reasons why X and Y are not independent. First, the next card is dealt from a deck that does not contain the two cards that you have already been dealt. Second, the sum Y includes X. **(b)** Since the two rolls are independent the two sums should be independent.

4.53 (a) The spins are independent. (b) Wrong. The probabilities are not equally likely because the deck now contains more black cards than red cards.

4.54 (a) 11+20=31. (b) No. (c) No.

4.55 40+25+5=70.

4.56 From Exercise 5.42, the mean is 11.251. The variance is $.010(4 - 11.251)^2 + .007(5-11.251)^2 + .007(6-11.251)^2 + .013(7-11.251)^2 + .032(8-11.251)^2 + .068(9-11.251)^2 + .070(10-11.251)^2 + .041(11-11.251)^2 + .752(12-11.251)^2 = 2.444$. The standard deviation is $\sqrt{2.444} = 1.563$.

4.57 The mean is $.03(0) + .16(1) + .30(2) + .23(3) + .17(4) + .11(5) = 2.68$. The variance is $.03(0 - 2.68)^2 + .16(1 - 2.68)^2 + .30(2 - 2.68)^2 + .23(3 - 2.68)^2 + .17(4 - 2.68)^2 + .11(5 - 2.68)^2 = 1.7177$. The standard deviation is $\sqrt{1.7177} = 1.3106$.

4.58 The mean for household is $1(.240)+2(.322)+3(.177)+4(.155)+5(.067)+6(.024)+7(.015) = 2.62$. The mean for family is $1(.000)+2(.413)+3(.236)+4(.211)+5(.090)+6(.032)+7(.018) = 3.15$. The variance of the number of people in a household is $.240(1-2.62)^2 + .322(2-2.62)^2 + .177(3-2.62)^2 + .155(4-2.62)^2 + .067(5-2.62)^2 + .024(6-2.62)^2 + .015(7-2.62)^2 = 2.02$. The standard deviation is $\sqrt{2.02} = 1.42$. The variance of the number of people in a family is $.413(2-3.15)^2 + .236(3-3.15)^2 + .211(4-3.15)^2 + .090(5-3.15)^2 + .032(6-3.15)^2 + .018(7-3.15)^2 = 1.54$. The standard deviation is $\sqrt{1.54} = 1.24$. The two distributions are very skewed to the right. It is difficult to judge which is more variable. The number of people in a household has the larger standard deviation. The standard deviation is not a good measure of variability for these distributions.

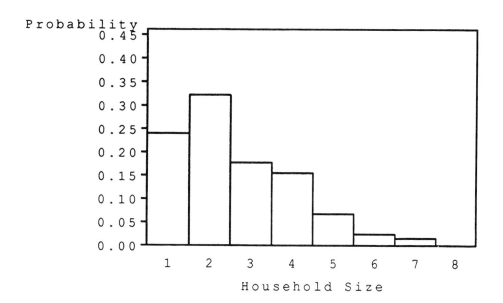

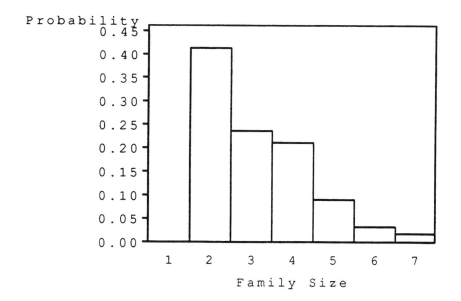

4.59 $\sigma^2_{X+Y+5} = \sigma^2_X + \sigma^2_Y = 2^2 + 1^2 = 5$. The standard deviation is $\sqrt{5} = 2.2361$.

4.60 $\sigma^2_{X+Y} = \sigma^2_X + \sigma^2_Y = 2^2 + 4^2 = 20$. The standard deviation is $\sqrt{20} = 4.47$.

4.61 (a) The variance of the number of civilian units is $.4(300 - 445)^2 + .5(500-445)^2 + .1(750-445)^2 = 19225$. The standard deviation is $\sqrt{19225} = 138.65$. **(b)** $\sigma^2_{X+Y} = \sigma^2_X + \sigma^2_Y = 7,800,000 + 19,225 = 7,819,225$. The standard deviation is $\sqrt{7,819,225} = 2796$. **(c)** $\sigma^2_Z = (2000)^2 \sigma^2_X + (3500)^2 \sigma^2_Y = 31,435,506,250,000$. The standard deviation is 5,606,738.

4.62 (a) The two scores are independent because the two students are randomly selected. **(b)** $\mu_{X-Y} = \mu_X - \mu_Y = 120 - 105 = 15$. $\sigma^2_{X-Y} = \sigma^2_X + \sigma^2_Y = 28^2 + 35^2 = 2009$. The standard deviation is $\sqrt{2009} = 44.82$. **(c)** No, the probability cannot be found because the distribution of $X - Y$ is not given.

4.63 (a) The mean temperature is $\mu_X = 540(.1) + 545(.25) + 550(.3) + 555(.25) + 560(.1) = 550$. The variance is $\sigma^2_X = (540-550)^2(.1) + (545-550)^2(.25) + (550-550)^2(.3) + (555-550)^2(.25) + (560-550)^2(.1) = 32.5$. The standard deviation is 5.7. **(b)** We subtract 550 from the mean so the mean number of degrees off target is 0. Subtracting a constant does not change the standard deviation. It is 5.7. **(c)** We use the rules for means and variances for $a + bX$. Here $a = 32$ and $b = 9/5$. The mean in Fahrenheit is $32 + (9/5)(550) = 1022$. The variance is $(9/5)^2 \sigma^2$, so the standard deviation is $(9/5)\sigma = (9/5)(5.7) = 10.3$.

4.64 (a) The mean of the difference is $\mu_Y - \mu_X = 2.001 - 2.000 = .001$. **(b)** First, we note that $\mu_{X+Y} = \mu_X + \mu_Y = 2.000 + 2.001 = 4.001$. Therefore, $\mu_Z = (1/2)\mu_{X+Y} = (1/2)(4.001) = 2.0005$. In a similar way we use the rules for variances in steps. The variance of the sum is $\sigma^2_{X+Y} = \sigma^2_X + \sigma^2_Y = (.002)^2 + (.001)^2 = .000004 + .000001 = .000005$. So, $\sigma^2_Z = (1/2)^2 \sigma^2_{X+Y} = (1/4)(.000005) = .00000125$. Therefore, $\sigma_Z = \sqrt{.00000125} = .0011$. The average is therefore slightly more variable than Y.

4.65 The variance is $\sigma^2_x = (-99,750 - 303.36)^2(.00183) + (-99,500 - 303.36)^2(.00186) + (-99,250 - 303.36)^2(.00189) + (-99,000 - 303.36)^2(.00191) + (-98,750 - 303.36)^2(.00193) + (1250 - 303.36)^2(.99058) = 76804994.25$. The standard deviation is $\sqrt{76804994.25} = 8763.85$.

4.66 (a) $\mu_{X_1+X_2} = 303.36 + 303.36 = 606.72$. $\sigma^2_{X_1+X_2} = 76804994.25 + 76804994.25 = 153609988.50$. The standard deviation is $\sqrt{153609988.50} = 12393.95$. **(b)** $\mu_Z = 1/2(\mu_{X_1} + \mu_{X_2}) = (1/2)(606.72) = 303.36$. $\sigma^2_Z =$

$(1/2)^2(\sigma_{X_1}^2 + \sigma_{X_2}^2) == (1/2)^2(\sigma_{X_1+X_2}^2 = (1/4)(153609988.50) = 38402497.12$. The standard deviation is $\sqrt{38402497.12} = 6196.97$. (c) $\mu_Z = 1/4(\mu_{X_1} + \mu_{X_2} + \mu_{X_3} + \mu_{X_4}) = (1/4)(4(303.36)) = 303.36$. $\sigma_Z^2 = (1/4)^2(\sigma_{X_1}^2 + \sigma_{X_2}^2 + \sigma_{X_3}^2 + \sigma_{X_4}^2) = (1/16)(4(76804994.25)) = 19201248.56$. The standard deviation is 4381.92. The mean is the same but the standard deviation is smaller (it has been divided by $\sqrt{2}$).

4.67 $.192 + .221 - .092 = .321$.

4.68 $.6 + .4 - .2 = .8$.

4.69 (a) .092, the household is prosperous and educated. (b) $.192 - .092 = .100$, the household is prosperous and not educated. (c) $.221 - .092 = .129$, the household is not prosperous and educated. (d) $1 - (.192 + .221 - .092) = .679$, the household is not prosperous and not educated.

4.70 (a) $P(A \cap B) = .2$. (b) $P(A \cap B^c) = P(A) - P(A \cap B) = .6 - .2 = .4$. (c) $P(A^c \cap B) = P(B) - P(A \cap B) = .4 - .2 = .2$. (d) $P((A \cup B)^c) = 1 - P(A \cup B) = 1 - .8 = .2$.

4.71 (a) $.01 + .30 + .13 + .09 = .53$. (b) $.09/(.04 + .09) = .6923$. (c) $.53 + .13 - .09 = .57$.

4.72 (a) $24226/30904 = .78$. (b) $18153/30904 = .59$. (c) $15518/24226 = .64$. (d) No. The two events are independent if P(victim is male and a firearm is used)=P(victim is male)P(a firearm is used). P(victim is male and a firearm is used)=15518/30904=.50. P(victim is male)P(a firearm is used)=(.78)(.59)=.46. Therefore, the two events are not independent.

4.73 (a) $16520/85834 = .1925$. (b) $7888/19007 = .4150$. (c) $7888/16520 = .4775$. (d) Not independent. If X and Y were independent then the probability in part (a) would be equal to the conditional probability in part(b). Note that the answers to (a) and (b) will differ slightly if the marginal totals in the exercise are used rather than summing the individual entries.

4.74 (a) .15. (b) .20.

4.75 $.45(.26)=.1170$.

4.76 $.8(.4)=.32$.

4.77 (a) $P(A) = .859$, $P(B|A) = .953$, $P(B|A^c) = .899$. (b)

```
                                          Employed
                                 .953 /
                    White---------------
                   /             .047 \
           .859 /                        Unemployed
              /
Labor Force
              \
           .141 \                         Employed
                 \               .899 /
                    Nonwhite------------
                                 .101 \
                                          Unemployed
```

(c) $.859(.953) = .8186$, $(1 - .859)(.899) = .1268$, $.8186 + .1268 = .9454$.

4.78 $.32 + .2(.6) = .44$.

4.79 $.8186/.9454 = .8659$.

4.80 $.75 + .25(.20) = .80$.

```
                                          Correct
                                 1.0 /
                    Knows Answer--------
                   /             0.0 \
           .75 /                         Incorrect
              /
Question
              \
           .25 \                          Correct
```

```
                            \             .20 /
                      Does Not Know-------
                                          .80 \
                                               Incorrect
```

4.81 $.4(.3) + .4(.9) + .2(.5) = .58$.

4.82 $.75/.80 = .9375$.

4.83 $P(Y > X) = 1/2$; $P(Y < .5 \cap Y > X) = 1/8$; $P(Y < .5 | Y > X) = (1/8)/(1/2) = 1/4$.

4.84 (a) A, A, A, B. (b) $(.8)^3(.2) = .1024$. (c) $P(X = x) = (.8)^{x-1}.2$.

4.85 $.10(.73) + .85(.76) = .719$. The surgery offers him a slightly better chance of achieving his goal.

```
                                             Moderate Activity
                                       .76 /
                      Survives Without-Complications
                    /                  .24 \
              .85 /                          No Moderate Activity
                /
Surgery -.05  Does Not Survive
                \
              .10 \                          Moderate Activity
                  \                    .73 /
                      Survives With Complications
                                       .27 \
                                             No Moderate Activity
```

4.86 The probability that building a plant is more profitable than contracting the production to Hong Kong is $.9(.36)(.95) + .9(1 - .36)(.3) + .1(0) + .1(1)(.1) = .4906$. Contracting the production to Hong Kong is more profitable.

4.87 $(1 + 18 + 120 + 270)/100,000 = .00409$. $5000(1/100000) + 200(18/100000) + 25(120/100000) + 20(270/100000) = .17$.

4.88 (a) $\mu_X = .1(1) + .2(1.5) + .4(2) + .2(4) + .1(10) = 3.00$. $\sigma_X^2 = .1(1-3)^2 + .2(1.5-3)^2 + .4(2-3)^2 + .2(4-3)^2 + .1(10-3)^2 = 6.35$. $\sigma_X = \sqrt{6.35} = 2.52$. **(b)** $\mu_Y = .9\mu_X - .2 = 2.50$. $\sigma_Y = .9\sigma_X = 2.27$.

4.89 (a) $1 - (1/10000 + 1/1000 + 1/100 + 1/20) = .9389$. **(b)** $\mu = .9389(0) + .0001(1000) + .001(200) + .01(50) + .05(10) = 1.3$. **(c)** $\sigma^2 = .9389(0-1.3)^2 + .0001(1000-1.3)^2 + .001(200-1.3)^2 + .01(50-1.3)^2 + .05(10-1.3)^2 = 168.31$. $\sigma = 12.97$.

4.90 (a) $P(T > 30) = P(Z > (30-25)/5) = P(Z > 1) = 1 - .8413 = .1587$. **(b)** $P(Z < 2.33) = .99$. Therefore, $2.33 = (t-25)/5$, or $t = 25 + 5(2.33) = 36.65$ seconds.

4.91 $Y = a + bX$ implies $\sigma_Y = b\sigma_X = 20b$. Therefore $b = .05$. $\mu_Y = a + b\mu_X = a + .05(1400)$. Therefore $a = -70$.

4.92 (a) $S = \{3, 4, 5, 6, \ldots, 18\}$. **(b)** $6/216 = .0278$. **(c)** $\mu_{X_i} = (1/6)1 + (1/6)2 + (1/6)3 + (1/6)4 + (1/6)5 + (1/6)6 = 3.5$. $\sigma_{X_i}^2 = (1/6)(1-3.5)^2 + (1/6)(2-3.5)^2 + (1/6)(3-3.5)^2 + (1/6)(4-3.5)^2 + (1/6)(5-3.5)^2 + (1/6)(6-3.5)^2 = 2.91725$. $\mu_X = 3(3.5) = 10.5$. $\sigma_X^2 = 3(2.91725) = 8.7518$. $\sigma_X = \sqrt{8.7518} = 2.96$.

4.93 (a) $\mu_Z = .5(.11) + .5(.02) = .065$. $\sigma_Z^2 = .5^2(.28)^2 + .5^2(.05)^2 = .020225$. $\sigma_Z = .14$. **(b)** For the general case, $\mu_Z = \alpha(.11) + (1-\alpha)(.02)$. $\sigma_Z^2 = \alpha^2(.28)^2 + (1-\alpha)^2(.05)^2$. (α, μ, σ): .55, .0695, .16; .60, .0740, .17; .65, .0785, .18; .70, .0830, .20; .75, .0875, .21; .80, .0920, .22; .85, .0965, .24; .90, .1010, .25; .95, .1055, .27, 1.00, .1100, .28.

4.94 Let the different types of sides be labeled A, B, C, D. The probability of rolling ABCD in that order is $.4(.4)(.1)(.1) = .0016$. There are 24 (4 factorial) orders (permutations) of these letters. Therefore, the probability is $.0016(24) = .0384$.

4.95 (a) and (b)

Ann Shows	Ann Guesses	Bob Shows	Bob Guesses	X
1	1	1	1	0
1	1	1	2	2
1	1	2	1	-3
1	1	2	2	0
2	1	1	1	3
2	1	1	2	0
2	1	2	1	0
2	1	2	2	-4
1	2	1	1	-2
1	2	1	2	0
1	2	2	1	0
1	2	2	2	3
2	2	1	1	0
2	2	1	2	-3
2	2	2	1	4
2	2	2	2	0

(c)

X	-4	-3	-2	0	2	3	4
freq	1	2	1	8	1	2	1
prob	.0625	.125	.0625	.5	.0625	.125	.0625

(d) $\mu_X = .0625(-4) + .125(-3) + .0625(-2) + .5(0) + .0625(2) + .125(3) + .0625(4) = 0$. $\sigma_X^2 = .0625(-4-0)^2 + .125(-3-0)^2 + .0625(-2-0)^2 + .5(0-0)^2 + .0625(2-0)^2 + .125(3-0)^2 + .0625(4-0)^2 = 4.75$. $\sigma_X = \sqrt{4.75} = 2.18$.

4.96 (a) $.057 + .010 = .067$. **(b)** $.01/.067 = .1493$.

4.97 $1 - (.14 + .03 - .01) = .84$.

4.98 (a) $.37 + .06 = .43$. **(b)** $.13(.37) = .0481$. **(c)** $2(.0481) = .0962$. **(d)** The probability that at least one has type O blood is the probability that the husband has type O blood plus the probability that the wife has type O blood and the husband does not. Thus, $.44 + .44(1 - .44) = .6864$. **(e)**

$.37^2 + .13^2 + .44^2 + .06^2 = .351$.

4.99 (a) $.11/(.11 + .14) = .44$. **(b)** $(.01 + .04)/(.01 + .04 + .11 + .12) = .18$.

4.100

```
                                            Yes
                                       .3 /
                      Heads --------------
                     /                 .7 \
                .5 /                        No
                 /
    Toss Coin
                 \
                .5 \                        Yes
                   \                 1.0 /
                      Tails --------------
                                       0.0 \
                                            No
```

$.7(.5) + 0(.5) = .35$. $.8(.5) + 0(.5) = .40$. $.39/.5 = .73$. We estimate that 1−.73 or 22% of the students have plagiarized a paper.

4.101 (a)

```
                                            Positive
                                       .997/
                   Antibodies Present---
                  /                    .003\
             .01 /                          Negative
               /
    Select a person
               \
             .99 \                          Positive
                 \                     .015/
                   Antibodies Absent----
                                       .985\
                                            Negative
```

(b) $.01(.997) + .99(.015) = .0249$. **(c)** $(.01)(.997)/.0249 = .4016$.

4.102 (a) We simulated the contents of 100 bags of 25 M&M's. The overall proportion of orange candies is $262/2500 = .1048$. The observed proportion is close to the probability .1. **(b)** The mean number of orange candies in the bags is $262/100 = 2.62$. It is close to the theoretical mean of 2.5. **(c)** The number of bags containing no orange candies is 6. We expect this to happen very infrequently. In our simulation it happened 6 times out of 100. The true probability of this happening is $(1-.1)^{25} = .072$.

4.103 (a) For 100 students there will be 5000 repetitions. The approximate probability distribution is $(x, p) = (0, .002), (1, .007), (2, .043), (3, .117), (4, .203), (5, .250), (6, .208), (7, .120), (8, .041), (9, .008), (10, .001)$.

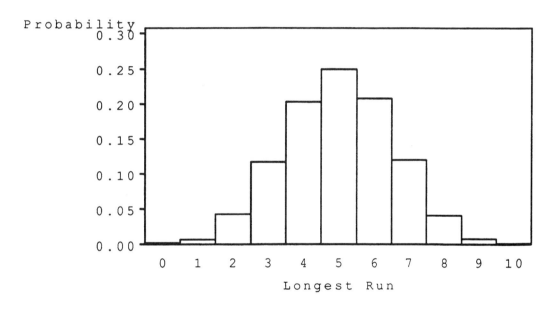

(b) The relative frequency of runs of 3 or more heads in our simulation is .96. **(c)** The mean computed from the probabilities in (a) is $\mu_X = 5.01$. This is the result obtained from 5000 repetitions. For one sample of 50 repetitions, the mean is 5.16. These values are close. Note than answers will vary with different simulations.

4.5 CHAPTER 5

5.1 (a) Yes. Let success be the event that a girl is born. Then X is the number of successes in 50 independent trials each having probability p of success. **(b)** No. n is random. **(c)** No. Trials are not independent.

5.2 (a) No. The probabilities for different types of defects are not the same. **(b)** Yes. Here n is 100 and p is the probability of saying yes. **(c)** Yes. Here n is 52 and p is the probability of winning in a given week.

5.3 (a) Yes. The sample size is 50 and p is the probability that a randomly selected student passes the exam. **(b)** No. Trials are not independent. **(c)** No. Each trial is performed under different conditions.

5.4 (a) $.0047 + .0305 + .0916 + .1700 = .2968$. **(b)** Ten or more whites is the same event as five or less blacks. $.2968 + .2186 + .2061 = .7215$.

5.5 (a) To do the problem using Table C, we work with X, the number of players who do not graduate. Here, $n = 20$ and $p = .2$. If 10 players graduate then 10 players do not graduate. Therefore, we find $P(X = 10) = .0020$. **(b)** If 10 or fewer players graduate, then 10 or more players do not graduate; so we calculate $P(X \geq 10) = .0020 + .0005 + .0001 + 0 = .0026$.

5.6 $n = 4$, $p = .25$.

x	0	1	2	3	4
probability	.3164	.4219	.2109	.0469	.0039

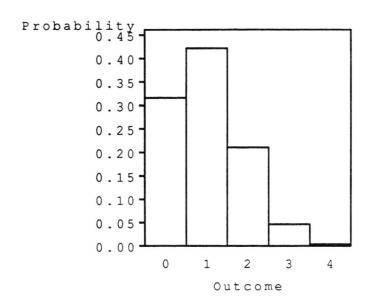

5.7 $n = 6$, $p = .65$. $P(X = 0) = .0018$, $P(X = 1) = .0205$, $P(X = 2) = .0951$, $P(X = 3) = .2355$, $P(X = 4) = .3280$, $P(X = 5) = .2437$, $P(X = 6) = .0754$.

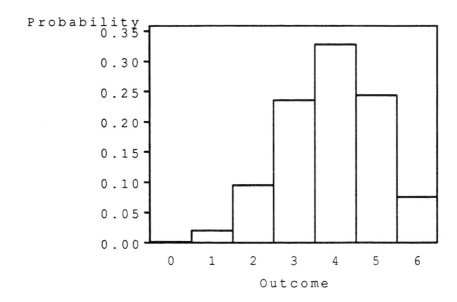

5.8 (a) $P(X = 2) = .2816$. **(b)** $P(X \leq 2) = .0563 + .1877 + .2816 = .5256$.

(c) This is a different description for the event described in part (b). .5256.

5.9 (a) There is only one way in which n successes can be chosen from among n trials. **(b)** There are n ways in which $n-1$ successes can be chosen from among n trials. **(c)** The number of ways in which k successes can be chosen from among n trials is the same as the number of ways in which $n-k$ successes can be chosen from among n trials.

5.10 $\mu = np = 4(1/4) = 1$. $\sigma = \sqrt{np(1-p)} = \sqrt{4(1/4)(3/4)} = .8660$.

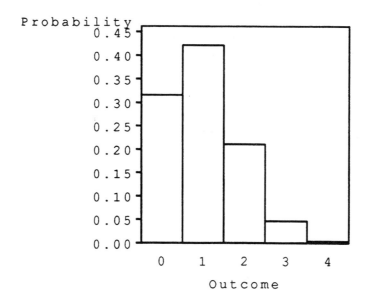

5.11 $\mu = np = 6(.65) = 3.9$. $\sigma = \sqrt{np(1-p)} = \sqrt{6(.65)(.35)} = 1.1683$.
$P(2.7317 \leq X \leq 5.0683) = P(3 \leq X \leq 5) = .2355 + .3280 + .2437 = .8072$.

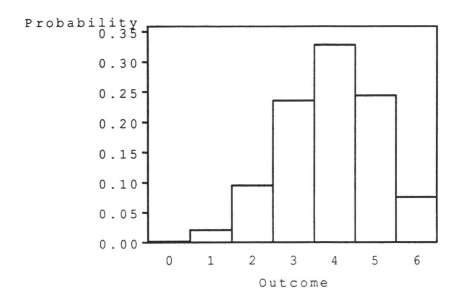

5.12 (a) .25. (b) X is B(20, .25). $P(X \geq 10) = .0099 + .0030 + .0008 + .0002 = .0139$. (c) $\mu = np = 20(1/4) = 5$. $\sigma = \sqrt{np(1-p)} = \sqrt{20(1/4)(3/4)} = 1.94$. (d) No, the trials are not independent.

5.13 (a) $p = .4$. $P(X \geq 3) = .2304 + .0768 + .0102 = .3174$. (b) $P(Y \geq 8) = .1181 + .0612 + .0245 + .0074 + .0016 + .0003 + 0 = .2131$.

5.14 (a) $P(X \geq 1) = 1 - (P(X = 0) + P(X = 1)) = 1 - (.0687 + .2062) = .7251$. (b) $\mu = np = 12(.2) = 2.4$. $\sigma = \sqrt{np(1-p)} = \sqrt{12(.2)(.8)} = 1.39$. (c) $P(X \leq 2) = .0687 + .2062 + .2835 = .5584$.

5.15 (a) $\mu = np = 300(.22) = 66$. $\sigma = \sqrt{np(1-p)} = \sqrt{300(.22)(.78)} = 7.17$. (b) $P(X \geq 80) = P(Z \geq (80 - 66)/7.17) = 1 - .9744 = .0256$, (with continuity correction, .0301).

5.16 (a) $\mu = np = 1500(.11) = 165$. $\sigma = \sqrt{np(1-p)} = \sqrt{1500(.11)(.89)} = 12.12$. (b) $P(X \leq 100) = P(Z \leq (100 - 165)/12.12) = P(Z \leq -5.36) = 0$. (with continuity correction, $P(X \leq 100) = P(Z \leq (100.5 - 165)/12.12) = P(Z \leq -5.32) = 0$.)

5.17 (a) $\mu = np = 1500(.7) = 1050$, $\sigma = \sqrt{np(1-p)} = \sqrt{1500(.7)(.3)} = 17.75$. (b) $P(X \geq 1000) = P(Z \geq (1000 - 1050)/17.75) = P(Z \geq -2.82) = 1 - .0024 = .9976$. (with continuity correction, $P(X \geq 1000) = P(Z \geq (999.5 - 1050)/17.75) = P(Z \geq -2.85) = 1 - .0022 = .9978$.) (c) $P(X > 1200) = P(Z > (1200 - 1050)/17.75) = P(Z > 8.45) = 1 - 1 = 0$. (d) $\mu = np = 1700(.7) = 1190$, $\sigma = \sqrt{np(1-p)} = \sqrt{1700(.7)(.3)} = 18.89$. $P(X > 1200) = P(Z > (1200 - 1190)/18.89) = P(Z > .53) = 1 - .7019 = .2981$. (with continuity correction, $P(X > 1200) = P(Z > (1200.5 - 1190)/18.89) = P(Z > .56) = 1 - .7123 = .2877$.)

5.18 (a) $\mu = p = .15$. $\sigma = \sqrt{p(1-p)/n} = \sqrt{.15(.85)/1540} = .0091$. (b) $P(.13 \leq \hat{p} \leq .17) = P(Z \leq (.17 - .15)/.0091) - P(Z \leq (.13 - .15)/.0091) = .9861 - .0139 = .9722$. (c) $\sigma(\hat{p})/2 = (1/2)\sqrt{.15(.85)/1540} = \sqrt{.15(.85)/n}$. Solving for n gives 6160.

5.19 (a) $\mu = 75$. $\sigma = 4.33$. $P(X \leq 70) = P(Z \leq (70 - 75)/4.33) = P(Z \leq -1.15) = .1251$. (with continuity correction, $P(X \leq 70) = P(Z \leq (70.5 - 75)/4.33) = P(Z \leq -1.04) = .1492$.) (b) $\mu = 187.5$. $\sigma = 6.8465$. $P(X \leq 175) = P(Z \leq (175 - 187.5)/6.8465) = P(Z \leq -1.83) = .0336$. (with continuity correction, $P(X \leq 175) = P(Z \leq (175.5 - 187.5)/6.8465) = P(Z \leq -1.75) = .0401$.) (c) To halve the standard deviation, we double the sample size. $n = 400$. (d) Yes.

5.20 (a) $\mu = 1000(1/5) = 200$. $\sigma = \sqrt{1000(1/5)(4/5)} = 12.65$. (b) $\mu = p = .20$. $\sigma = \sqrt{.2(.8)/1000} = .0126$. (c) $P(\hat{p} \geq .24) = P(Z \geq (.24 - .20)/.0126) = P(Z \geq 3.17) = 1 - .9992 = .0008$. (d) $z = 2.33 = (p - .20)/.0126$. $p = .2294$.

5.21 Let Y be the number that escape detection. Y is $B(20, .01)$. (a) From Table C, $P(Y = 0) = .8179$. (b) $P(Y > 1) = P(Y = 2) + P(Y = 3) + \cdots + P(Y = 20) = .0159 + .0010 + 0 = .0169$. (c) Let X be the number that are detected. $\mu_X = 20(.99) = 19.8$. $\sigma_X = \sqrt{20(.99)(.01)} = .445$.

5.22 Let X be the number of positives in the uncontaminated blood. Let Y be the number of positives in the contaminated blood. (a) The mean

CHAPTER 5   205

number of false positives is $\mu_X = 12000(.02) = 240$. The standard deviation is $\sigma_X = \sqrt{12000(.02)(.98)} = 15.34$. (b) The mean number of positives for the contaminated units is $\mu_Y = 20(.99) = 19.8$. The standard deviation is $\sigma_Y = \sqrt{20(.99)(.01)} = .445$. Let $Z = X + Y$ be the total number of positives. Then, the mean is $\mu_Z = \mu_X + \mu_Y = 240 + 19.8 = 259.8$ and the variance is $\sigma_Z^2 = \sigma_X^2 + \sigma_Y^2 = 15.34^2 + .445^2 = 235.40$. The standard deviation is $\sigma_Z = \sqrt{235.40} = 15.34$. (c) $\mu_X/\mu_Z = 240/259.8 = .9238 = 92.38\%$.

5.23 $\sigma(\bar{x}) = .08/\sqrt{3} = .0462$.

5.24 $\mu(\bar{x}) = \mu = 40.125$. $\sigma(\bar{x}) = .002/\sqrt{4} = .001$.

5.25 $\mu(\bar{x}) = \mu = 18.6$. $\sigma(\bar{x}) = 5.9/\sqrt{76} = .6768$.

5.26 $\mu(\bar{x}) = \mu = -3.5$. $\sigma(\bar{x}) = .26/\sqrt{5} = .1163$.

5.27 (a) $P(X \geq 21) = P(Z \geq (21 - 18.6)/5.9) = 1 - P(Z \leq .41) = 1 - .6591 = .3409$. (b) $\bar{x}$ is $N(18.6, 5.9/\sqrt{76}) = N(18.6, .6768)$. $P(\bar{x} \geq 20.4) = P(Z \geq (20.4 - 18.6)/.6768) = 1 - P(Z \leq 2.66) = 1 - .9961 = .0039$.

5.28 (a) $P(X < 295) = P(Z < (295 - 298)/3) = P(Z < -1) = .1587$. (b) $\mu(\bar{x}) = \mu_X = 298$. $\sigma(\bar{x}) = 3/\sqrt{6} = 1.22$. $P(\bar{x} < 295) = P(Z < (295 - 298)/1.22) = .0069$.

5.29 (a) $P(X < 3.5) = P(Z < (3.5 - 3.8)/.2) = .0668$. (b) $\mu(\bar{x}) = \mu_X = 3.8$. $\sigma(\bar{x}) = .2/\sqrt{4} = .1$. $P(\bar{x} < 3.5) = P(Z < (3.5 - 3.8)/.1) = .0013$.

5.30 (a) $\bar{x}$ is $N(123, .08/\sqrt{3}) = N(123, .05)$. (b) $P(\bar{x} \geq 124) = P(Z \geq (124 - 123)/.05) = 0$.

5.31 (a) $\bar{x}$ is $N(55000, 4500/\sqrt{8}) = N(55000, 1591)$. (b) $P(\bar{x} \leq 51,800) = P(Z \leq (51800 - 55000)/1591) = .0222$.

5.32 (a) Let X be $N(100, 2.5)$ and Y be $N(250, 2.8)$. $\mu_{X+Y} = \mu_X + \mu_Y = 100 + 250 = 350$. $\sigma_{X+Y}^2 = \sigma_X^2 + \sigma_Y^2 = 2.5^2 + 2.8^2 = 14.09$. $\sigma_{X+Y} = \sqrt{14.09} = 3.75$. Therefore, $X + Y$ is $N(350, 3.75)$. (b) $P(345 \leq X + Y \leq 355) =$

$P((345 - 350)/3.75 \leq Z \leq (355 - 350)/3.75) = P(Z \leq 1.33) - P(Z \leq -1.33) = .9082 - .0918 = .8164.$

5.33 (a) $\mu_{Y-X} = \mu_Y - \mu_X = .526 - .525 = .001.$ $\sigma^2_{Y-X} = \sigma^2_Y + \sigma^2_X = .0004^2 + .0003^2 = .00000025.$ $\sigma_{Y-X} = \sqrt{.00000025} = .0005.$ Therefore $Y - X$ is $N(.001, .0005)$. **(b)** $P(Y - X \leq 0) = P(Z \leq (0 - .001)/.0005) = .0228.$

5.34 (a) $\mu(\bar{x}) = \mu_X = 360.$ $\sigma(\bar{x}) = 55/\sqrt{20} = 12.2984.$ $\mu(\bar{y}) = \mu_Y = 385.$ $\sigma(\bar{y}) = 50/\sqrt{20} = 11.1803.$ $\mu(\bar{y} - \bar{x}) = \mu(\bar{y}) - \mu(\bar{x}) = 385 - 360 = 25.$ $\sigma^2(\bar{y} - \bar{x}) = \sigma^2(\bar{y}) + \sigma^2(\bar{x}) = 12.2984^2 + 11.1803^2 = 276.25.$ $\sigma(\bar{y} - \bar{x}) = \sqrt{276.25} = 16.6208.$ **(b)** $\bar{x}$ is $N(360, 12.2984)$. $\bar{y}$ is $N(385, 11.1803)$. $\bar{y} - \bar{x}$ is $N(25, 16.6208)$. **(c)** $P(\bar{y} - \bar{x} \geq 25) = P(Z \geq (25 - 25)/16.6208) = .5.$

5.35 (a) $\bar{x}$ is $N(34, 12/\sqrt{26}) = N(34, 2.3534).$ **(b)** $\bar{y}$ is $N(37, 11/\sqrt{24}) = N(37, 2.2454).$ **(c)** $\mu(\bar{y} - \bar{x}) = \mu(\bar{y}) - \mu(\bar{x}) = 37 - 34 = 3.$ $\sigma^2(\bar{y} - \bar{x}) = \sigma^2(\bar{y}) + \sigma^2(\bar{x}) = 2.3534^2 + 2.2454^2 = 10.5803.$ $\sigma(\bar{y} - \bar{x}) = \sqrt{10.5803} = 3.2527.$ Therefore, $\bar{y} - \bar{x}$ is $N(3, 3.2527)$. **(d)** $P(\bar{y} - \bar{x} \geq 4) = P(Z \geq (4-3)/3.2527) = P(Z \geq .31) = 1 - .6217 = .3783.$

5.36 (a) $\bar{y}$ is $N(\mu_Y, \sigma_Y/\sqrt{m})$. $\bar{x}$ is $N(\mu_X, \sigma_X/\sqrt{n})$. **(b)** $\mu(\bar{y} - \bar{x}) = \mu(\bar{y}) - \mu(\bar{x}) = \mu_Y - \mu_X.$ $\sigma^2(\bar{y} - \bar{x}) = \sigma^2(\bar{y}) + \sigma^2(\bar{x}) = \sigma^2_Y/m + \sigma^2_X/n.$ Therefore, $\bar{y} - \bar{x}$ is $N(\mu_Y - \mu_X, \sqrt{\sigma^2_Y/m + \sigma^2_X/n})$.

5.37 (a) $d_1 = 2(\sigma_X) = .004$, $d_2 = 2(\sigma_Y) = .002.$ **(b)** $\sigma^2_{X+Y+Z} = \sigma^2_X + \sigma^2_Y + \sigma^2_Z = .002^2 + .001^2 + .001^2 = i.000006.$ $\sigma_{X+Y+Z} = \sqrt{.000006} = .0024.$ $2(\sigma_{X+Y+Z}) = 2(.0024) = .0048.$ From part(a), $d_1 + 2d_2 = .004 + 2(.002) = .008$, No.

5.38 $\mu_{X-Y} = \mu_X - \mu_Y = 0.$ $\sigma_{X-Y} = \sqrt{\sigma^2_X + \sigma^2_Y} = \sqrt{.18} = .4243.$ $P(X - Y \geq .08) = P(Z \geq (.08 - 0)/.4243) = 1 - .5753 = .4247.$ $P(X - Y \leq -.08) = P(Z \leq (-.08 - 0)/.4243) = .4247.$ The sum of these is .8494.

5.39 $\mu_{X-Y} = \mu_X - \mu_Y = 0.$ $\sigma_{X-Y} = \sqrt{\sigma^2_X + \sigma^2_Y} = \sqrt{2^2 + 2^2} = 2.8284.$ $P(|X - Y| \geq 5) = 2P(Z \geq (5 - 0)/2.8284) = 2(1 - P(Z \leq 1.77)) = 2(1 - .9616) = .0768.$

5.40 (a) $\mu_{X+Y} = \mu_X + \mu_Y = 25 + 25 = 50$. $\sigma_{X+Y} = \sqrt{\sigma_X^2 + \sigma_Y^2} = \sqrt{10^2 + 9^2} = 13.45$. Therefore, $X + Y$ is $N(50, 13.45)$. **(b)** $P(X + Y \geq 60) = P(Z \geq (60 - 50)/13.45) = P(Z \geq .74) = 1 - .7704 = .23$. **(c)** Yes, the addition rule for means applies. If the scores are correlated the addition rule for variances used in part (a) does not apply.

5.41 (a) Yes, by the rules for means. **(b)** No, by the rules for variances. Variances add only if the two variables are uncorrelated.

5.42 $\bar{x}$ is $N(1.6, 1.2/\sqrt{200})$. $P(\bar{x} \geq 2) = P(Z \geq (2 - 1.6)/.0849) = 0$.

5.43 (a) $\bar{x}$ is $N(2.2, 1.4/\sqrt{52}) = N(2.2, .1941)$. **(b)** $P(\bar{x} < 2) = P(Z < (2 - 2.2)/.1941) = .1515$. **(c)** $P(\bar{x} < 100/52) = P(Z < (1.9231 - 2.2)/.1941) = .0764$.

5.44 $\bar{x}$ is $N(.09, .28/\sqrt{45})$. $P(\bar{x} > .15) = P(Z > (.15 - .09)/.0417) = 1 - P(Z < 1.44) = .0749$. $P(\bar{x} < .05) = P(Z < (.05 - .09)/.0417) = P(Z < -.96) = .1685$.

5.45 $\bar{x}$ is $N(1.4, .3/\sqrt{125})$. $P(Z > 2.326) = .01$. Therefore, $(L - 1.4)/.0268 = 2.326$ or $L = 1.46$.

5.46 $\bar{x}$ is $N(13.6, 3.1/\sqrt{22})$. $P(Z < -1.645) = .05$. Therefore, $(L - 13.6)/.66 = -1.645$ or $L = 12.5$.

5.47 $\mu(\bar{x}) = \mu_X = 75$. $\sigma(\bar{x}) = .5/\sqrt{4} = .25$. Center line $= 75$. Control limits are $75 + 3(.25) = 75.75$ and $75 - 3(.25) = 74.25$.

5.48 $\mu(\bar{x}) = \mu_X = .875$. $\sigma(\bar{x}) = .0012/\sqrt{5} = .00054$. Center line $= .875$. Control limits are $.875 + 3(.00054) = .87661$ and $.875 - 3(.00054) = .87339$.

5.49 $\mu(\bar{x}) = \mu_X = 11.5$. $\sigma(\bar{x}) = .2/\sqrt{4} = .1$. **(a)** Center line $= 11.5$. Control limits are $11.5 + 3(.1) = 11.80$ and $11.5 - 3(.1) = 11.20$. **(b)**

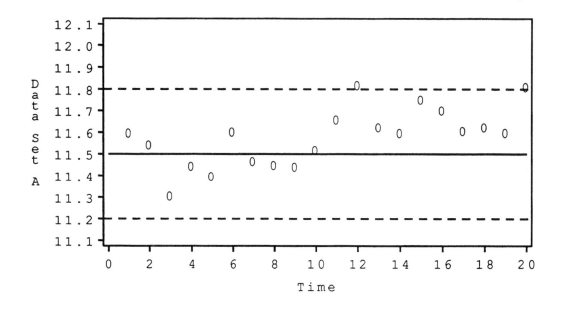

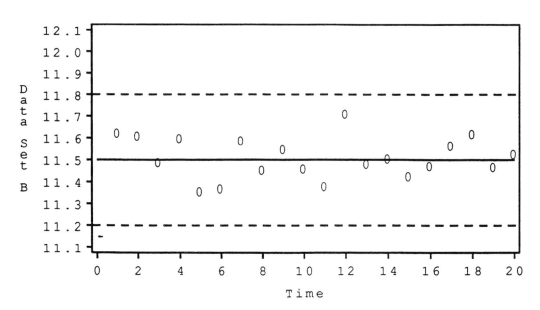

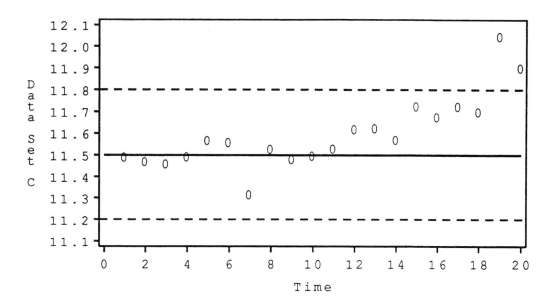

(c) Process B is in control. Process C shifts suddenly at time 11 or 12. Process A shifts gradually.

5.50 The center line is 2.205. The control limits are $2.205 \pm 3(.0010/\sqrt{5}) =$ 2.20366 and 2.20634. The one point out rule signals at hour 7. The run of nine rule gives no signal.

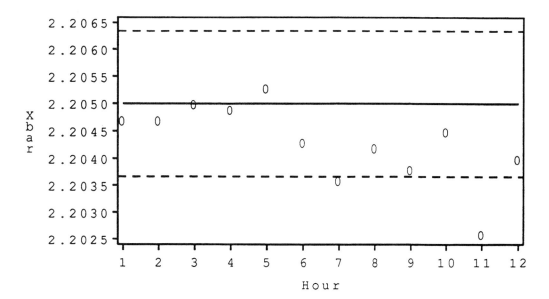

5.51 (a) Center line = 10. Control limits are $10 \pm 3(1.2/\sqrt{3}) = 12.08$ and 7.92.

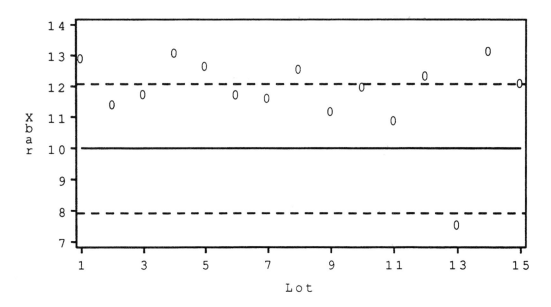

(b) Lot 13 is low and all of the other points are above the center line. Therefore, process is out of control. The high points do not call for remedial action but the low point signals a problem.

5.52 If the process is in control then each observation is described by the same distribution. In other words, the distribution is stable over time. Control charts focus on the process rather than on the products. They do not ensure good quality, but allow you to check the process at regular intervals to determine if the distribution has shifted. If the distribution has shifted, you can then correct the process.

5.53 The probability that a point is above $\mu + \sigma/\sqrt{n}$ is $1/2(1 - .68) = .16$. The probability that at least four of five are above $\mu + \sigma/\sqrt{n}$ is $P(X \geq 4)$ where X is $B(5, .16)$. $P(X \geq 4) = P(X = 4) + P(X = 5) = 5(.16)^4(.84) + .16^5 = .0029$. Therefore, the probability of an out of control signal is $2(.0029) = .0058$.

5.54 The probability that a point is within one sigma is .68. Therefore, the probability that the next 15 points are within one sigma is $.68^{15} = .003074$.

5.55 From Table A the following would be correct: 3.08, 3.09, and 3.10.

5.56 Center line = 162. Control limits are $162 \pm 2(1.3) = 159.4$ and 164.6. Joe's weight is out of control for six of the first eight weeks. It is systematically decreasing toward the control value.

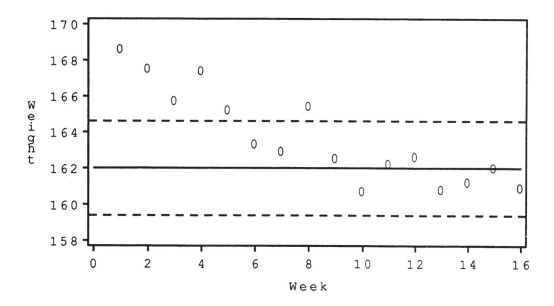

5.57 (a) $barx = 8.40$, $s = .62$. (b) Center line is at 8.40, the lower and upper control limits are at 7.15 and 9.65.

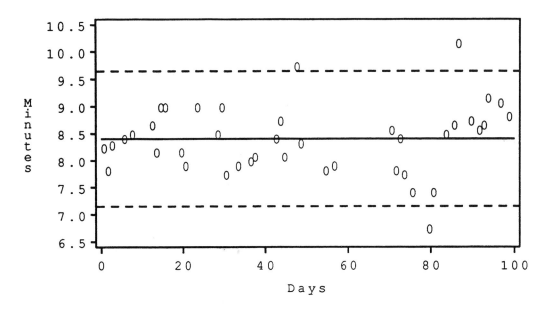

(c) Three points are out of control. The high values on October 27 and December 5 are explained by the truck and the ice as described in the problem. There is a low value on November 28. This was the day after Thanksgiving and there was little traffic. The last 10 points are all above the center line, probably due to bad winter weather.

5.58 (a) $\mu(\hat{p}) = p = .1$. $\sigma(\hat{p}) = \sqrt{p(1-p)/n} = \sqrt{.1(.9)/400} = .015$. **(b)** $N(.1, .015)$. **(c)** Center line $= .1$. Control limits are $.1 \pm 3(.015) = .055$ and $.145$. **(d)**

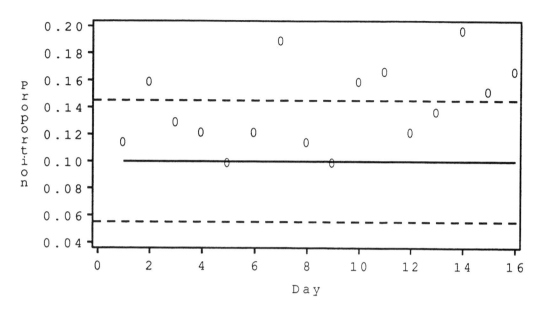

5.59 Center line $= p$. Control limits are $p \pm 3\sqrt{p(1-p)/n}$.

5.60 Center line $= .0225$. Control limits are $.0225 \pm 3\sqrt{.0225(.9775)/80} = -.0272$ and $.0522$. The inaccuracy of the normal approximation for this problem gives a negative lower control limit. This value should be replaced by 0.

5.61 Let X be the number of free throws made. X is $B(6, .7)$. $P(X \leq 2) = P(X = 0) + P(X = 1) + P(X = 2) = .07047$, using the computer. To use Table C we let Y be the number of free throws missed. Y is $B(6, .3)$. $P(Y \geq 4) = P(Y = 4) + P(Y = 5) + P(Y = 6) = .0595 + .0102 + .0007 = .0704$. Since Ray would make two or fewer free throws in six attempts about 7% of the time, we think that his performance is due to chance variation.

5.62 (a) $P(X \geq 105) = P(Z \geq (105 - 100)/15) = 1 - .6293 = .3707$. (b) $\mu(\bar{x}) = \mu = 100$. $\sigma(\bar{x}) = \sigma/\sqrt{n} = 15/\sqrt{60} = 12.91$. (c) $P(\bar{x} \geq 105) = P(Z \geq (105 - 100)/12.91) = 1 - .6519 = .3483$. (d) The answer in part (a) would change since the distribution is not normal. Parts (b) and (c) are based on the mean of a sample of size 60. The central limit theorem states that means are approximately normal. Therefore, the answers to parts (b) and (c) should be reasonably accurate.

5.63 (a) $\mu = np = 25000(.141) = 3525$. $\sigma = \sqrt{np(1-p)} = \sqrt{25000(.141)(.859)} = 55$. **(b)** $P(X \geq 3500) = P(Z \geq (3500-3525)/55) = 1 - .3264 = .6736$.

5.64 No, you are collecting signatures until you reach a total of 100 so there is not a fixed number of trials.

5.65 (a) To use Table C we need to use Y, the number of plants with white blossoms. Y is $B(8, .25)$. $P(Y = 2) = .3115$. **(b)** $\mu = np = 80(.75) = 60$. **(c)** $P(X \geq 50) = P(Z \geq (50-60)/3.8730) = 1 - .0049 = .9951$. (with continuity correction, $P(X \geq 50) = P(Z \geq (49.5-60)/3.8730) = 1 - .0034 = .9966$.)

5.66 X is $N(65, 5)$. Let Y equal the sum of the weights of the 12 eggs. $\mu_Y = \mu_{X_1} + \mu_{X_2} + \cdots + \mu_{X_{12}} = 12(65) = 780$. Similarly, $\sigma_Y^2 = \sigma_{X_1}^2 + \sigma_{X_2}^2 + \cdots + \sigma_{X_{12}}^2 = 12(25) = 300$. $\sigma_Y = \sqrt{300} = 17.32$. $P(750 \leq Y \leq 825) = P((750-780)/17.32 \leq Z \leq (825-780)/17.32) = P(Z \leq 2.60) - P(Z \leq -1.73) = .9953 - .0418 = .9535$.

5.67 (a) $B(500, .52)$. **(b)** $\mu = np = 500(.52) = 260$. $\sigma = \sqrt{np(1-p)} = \sqrt{500(.52)(.48)} = 11.17$. $P(X \geq 250) = P(Z \geq (250-260)/11.17) = 1 - .1841 = .8159$. (with continuity correction, $P(X \geq 250) = P(Z \geq (249.5-260)/11.17) = 1 - .1736 = .8264$.

5.68 (a) No, there is not a fixed number of trials with the same probability of success on each trial. **(b)** No, X cannot be negative and only whole-number values are possible. **(c)** $\mu(\bar{x}) = \mu_X = 1.5$. $\sigma(\bar{x}) = \sigma/\sqrt{n} = .75/\sqrt{700} = .0283$. Using the central limit theorem, $\bar{x}$ is $N(1.5, .02835)$. **(d)** Let $y = 700\bar{x}$. $\mu(y) = 700\mu(\bar{x}) = 700(1.5) = 1050$. $\sigma^2(y) = 700^2\sigma^2(\bar{x}) = 394$. $\sigma(y) = \sqrt{394} = 19.8$. Therefore, y is $N(1050, 19.8)$. $P(y > 1075) = P(Z > (1075-1050)/19.8) = 1 - .8962 = .1038$. (with continuity correction, $P(y > 1075) = P(Z > (1075.5-1050)/19.8) = 1 - .9015 = .0985$.

5.69 Let X equal the number in favor of vouchers. X is $B(500, .45)$. $\mu = np = 500(.45) = 225$. $\sigma = \sqrt{np(1-p)} = \sqrt{500(.45)(.55)} = 11.1243$.

$P(X \geq 250) = P(Z \geq (250 - 225)/11.1243) = 1 - .9878 = .0122.$ (with continuity correction, $P(X \geq 250) = P(Z \geq (249.5 - 225)/11.1243) = 1 - .9861 = .0139.$)

5.70 (a) The machine that fastens the cap and the machine that applies the torque are not the same. (b) Let Y equal the torque minus the strength $(T - S)$. $\mu_Y = \mu_T - \mu_S = 7 - 10 = -3$. $\sigma_Y^2 = \sigma_T^2 + \sigma_S^2 = .9^2 + 1.2^2 = 2.25$. $\sigma_Y = 1.5$. $P(Y > 0) = P(Z > (0 - (-3))/1.5) = 1 - .9772 = .0228$.

5.71 Center line $= 10$. Control limits are $10 \pm 3(1.2/\sqrt{6}) = 8.53$ and 11.47.

5.72 (a) $P(X < 2.8) + P(X > 3.2) = P(Z < (2.8 - 2.829)/.1516) + P(Z > (3.2 - 2.829)/.1516) = P(Z < -.19) + P(Z > 2.45) = .4247 + (1 - .9929) = .4318$. (b) Center line $= 3.0$. Control limits are $3.0 \pm 3(.1516/\sqrt{5}) = 2.797$ and 3.203.

5.73 (a) $\mu(\bar{x}) = 32$. $\sigma(\bar{x}) = \sigma_X/\sqrt{n} = 6/\sqrt{23} = 1.25$. $\bar{x}$ is $N(32, 1.25)$, $\mu(\bar{y}) = 29$. $\sigma(\bar{y}) = \sigma_Y/\sqrt{n} = 5/\sqrt{23} = 1.04$. $\bar{y}$ is $N(29, 1.04)$. (b) Let $D = \bar{y} - \bar{x}$. $\mu(D) = \mu(\bar{y}) - \mu(\bar{x}) = 29 - 32 = -3$. $\sigma^2(D) = \sigma^2(\bar{y}) + \sigma^2(\bar{x}) = 1.04^2 + 1.25^2 = 2.644$. $\sigma(D) = \sqrt{2.644} = 1.63$. Therefore, D is $N(-3, 1.63)$. (c) $P(\bar{y} \geq \bar{x}) = P(D \geq 0) = P(Z \geq (0 - (-3))/1.63) = 1 - .9671 = .0329$.

5.74 (a) There are 900 observations generated. It is impractical for students to list these. (b) There are 100 values of $\bar{x}$. These can be listed and will vary from student to student. (c) In our simulation the mean was 65.4475 and the standard deviation was .8524. The distribution looks approximately normal. The theoretical distribution for comparison is the normal distribution with mean 65.5 and standard deviation $\sigma = 2.5/\sqrt{3} = .8333$.

5.75 Answer will vary with the software used. SAS IRCHART uses control limits at 7.4 and 9.4. In Exercise 5.57, we found the center line to be at $\bar{x} = 8.4$ and control limits at 7.15 and 9.65. In general, the more complicated methods of calculating the standard deviation will result in limits that are closer to the center line.

4.6 CHAPTER 6

6.1 (a) Ninety-five percent of all samples we could select would give intervals that capture the true percentage of people who intend to vote for Carter. **(b)** The interval is 49% to 53%. This includes the value 50%, so a clear winner is not evident. **(c)** Confidence intervals do not give this kind of result. Either half of the voters favor Carter or not. There is no probability associated with the truth or falsehood of this unknown fact.

6.2 (a) This particular poll gave the result 47%. It is based on a sample and not all women were included. We expect the true percentage to be within 3 percentage points (the margin of error) of the poll value. **(b)** Ninety-five percent of all samples we could select would give intervals that capture the true percentage of women who would say that they do not get enough time for themselves. **(c)** The margin of error for men is larger because the percentage for men is based on a smaller sample than the percentage for women.

6.3 No. The confidence interval attempts to capture the true mean, not a proportion of the population.

6.4 The result is not trustworthy because the callers are not a random sample of citizens. This is an example of voluntary response.

6.5 (a) $\sigma(\bar{x}) = 5/\sqrt{24} = 1.02$. **(b)** $\bar{x} \pm z^*\sigma/\sqrt{n} = 60 \pm 1.960(1.02) = (58.00, 62.00)$. We are quite sure that the average weight of the population of runners is less than 65 kg.

6.6 (a) $\bar{x} \pm z^*\sigma/\sqrt{n} = 130 \pm 1.645(10/\sqrt{50}) = (127.67, 132.33)$. **(b)** $\bar{x} \pm z^*\sigma/\sqrt{n} = 130 \pm 1.960(10/\sqrt{50}) = (127.23, 132.77)$. **(c)** $\bar{x} \pm z^*\sigma/\sqrt{n} = 130 \pm 2.576(10/\sqrt{50}) = (126.36, 133.64)$. **(d)** As the confidence level increases, the margin of error increases.

6.7 $\bar{x} \pm z^*\sigma/\sqrt{n} = 60 \pm 2.576(1.02) = (57.37, 62.63)$. The 99% confidence interval is wider. More confidence requires a larger interval.

6.8 (a) $\bar{x} \pm z^*\sigma/\sqrt{n} = 130 \pm 1.960(10/\sqrt{100}) = 130 \pm 1.96 = (128.04, 131.96)$. **(b)** The margin of error is 1.96. The margin of error for the sample of size 50 is $z^*\sigma/\sqrt{n} = 1.96(10)/\sqrt{50} = 2.77$. The interval for the larger sample size is smaller because the standard deviation of the mean is smaller. (c) For any confidence level the larger sample size will give a smaller interval.

6.9 (a) $\bar{x} \pm z^*\sigma/\sqrt{n} = 3.2 \pm 1.645(.2/\sqrt{1}) = (2.87, 3.53)$. (b) $\bar{x} \pm z^*\sigma/\sqrt{n} = 3.2 \pm 1.645(.2/\sqrt{3}) = (3.01, 3.39)$.

6.10 $\bar{x} \pm z^*\sigma/\sqrt{n} = 2.13 \pm 1.960(.8/\sqrt{42}) = (1.89, 2.37)$.

6.11 $\bar{x} \pm z^*\sigma/\sqrt{n} = 224 \pm 1.960(.06/\sqrt{16}) = (223.973, 224.031)$.

6.12 $\bar{x} \pm z^*\sigma/\sqrt{n} = 35.09 \pm 2.576(11/\sqrt{44}) = (30.82, 39.36)$.

6.13 (a) $m = z^*\sigma_X/\sqrt{n} = 1.960(12/\sqrt{100}) = 2.35$. (b) $m = z^*\sigma_X/\sqrt{n} = 1.960(12/\sqrt{10}) = 7.44$. (c) $n = (z^*\sigma/m)^2 = (1.960(12)/5)^2 = 22.13$. A sample of size 23 is needed. Yes, the budget allows as many as 100 students.

6.14 $n = (z^*\sigma/m)^2 = (1.960(.06)/.02)^2 = 34.57$. A sample of size 35 is needed.

6.15 $n = (z^*\sigma/m)^2 = (1.645(10)/5)^2 = 10.82$. A sample of size 11 is needed.

6.16 (a) $\bar{x} \pm z^*\sigma/\sqrt{n} = 10.0023 \pm 2.326(.0002/\sqrt{5}) = (10.0021, 10.0025)$.
(b) $n = (z^*\sigma/m)^2 = (2.326(.0002)/.0001)^2 = 21.64$. A sample of size 22 is needed.

6.17 $\bar{x} \pm z^*\sigma/\sqrt{n} = 23453 \pm 2.576(8721/\sqrt{2621}) = (23015, 23891)$.

6.18 (a) No we are 95% confident that the true population percent falls in this interval. (b) $30 \pm 3 = (27, 33)$. (c) If $z^*\sigma_{estimate} = 3$, then $\sigma_{estimate} = 3/z^* = 3/1.960 = 1.5306$. (d) No.

6.19 $estimate \pm z^*\sigma_{estimate} = 664 \pm 1.960(3.50) = (657.14, 670.86)$.

6.20 (a) The probability is $.95^7 = .6983$. (b) The number of intervals (X) that cover their parameters is $B(7, .95)$. $P(X \geq 6) = P(X = 6) + P(X = 7) = 7(.95)^6(.05) + .95^7 = .9556$.

6.21 (a) $H_0: \mu = 1250$, $H_a: \mu < 1250$. (b) $H_0: \mu = 32$, $H_a: \mu > 32$. (c) $H_0: \mu = 5$, $H_a: \mu \neq 5$.

6.22 (a) $H_0: \mu = 18$, $H_a: \mu < 18$. (b) $H_0: \mu = 50$, $H_a: \mu > 50$. (c) $H_0: \mu = 24$, $H_a: \mu \neq 24$.

6.23 (a) $H_0: p_1 = p_2$, $H_a: p_1 > p_2$, where p_1 is the percent of males who name mathematics as their favorite and p_2 is the percent of females. (b) $H_0: \mu_A = \mu_B$, $H_a: \mu_A > \mu_B$, where μ_A is the mean score for group A and μ_B is the mean score for group B. (c) $H_0: \rho = 0$, $H_a: \rho > 0$, where ρ is the correlation between income and the percent of disposable income that is saved.

6.24 If the average blood pressures in the two groups are equal, the probability of observing a difference as extreme as that actually observed is .008. Since the P-value is small, we conclude that the means are different.

6.25 If we assume that the percentage of church attenders who are ethnocentric is equal to the percentage of nonattenders who are ethnocentric, then the probability of observing a difference as extreme as that actually observed is less than .05.

6.26 If we assume that the two sexes had the same average earnings, then the chance of observing a difference as extreme as that actually observed is .038. For race the corresponding probability is .476. Since the probability is small for sex and large for race we conclude that there is evidence for a sex difference but no clear evidence for a race difference.

6.27 $z = (\bar{x} - \mu_0)/(\sigma/\sqrt{n}) = (8.2 - 6.9)/(2.7/\sqrt{5}) = 1.07$. $P = P(Z > 1.07) = .1423$. There is no evidence that the new sonnets are not by the poet.

6.28 (a) $z = (\bar{x} - \mu_0)/(\sigma/\sqrt{n}) = (125.2 - 115)/(30/\sqrt{25}) = 1.70$. $P = P(Z > 1.70) = 1 - .9554 = .0446$. There is evidence to support the claim that older students have better attitudes toward school. (b) An SRS selected from a population with a normal distribution. The assumption that we have an SRS is more important.

6.29 $z = (\bar{x} - \mu_0)/(\sigma/\sqrt{n}) = (123.6 - 120)/(10/\sqrt{50}) = 2.55$. $P = 2P(Z > 2.55) = 2(1 - .9946) = .0108$. There is evidence to conclude that the population mean is not 120 bushels per acre. The normal assumption is not critical because the sample size is 50.

6.30 (a) $H_0 : \mu = 20$, $H_a : \mu > 20$. $z = (\bar{x} - \mu_0)/(\sigma/\sqrt{n}) = (21.1 - 20)/(6/\sqrt{43}) = 1.20$. $P = P(Z > 1.20) = 1 - .8849 = .1151$. The data do not provide sufficient evidence to conclude that the new course has improved the students' ACT scores. **(b)** Randomly assign students to two groups; one group will take the course and the other will not.

6.31 (a) $H_0 : \mu = 224$, $H_a : \mu \neq 224$. **(b)** $z = (\bar{x} - \mu_0)/(\sigma/\sqrt{n}) = (224.002 - 224)/(.06/\sqrt{16}) = .13$. $P = 2P(Z > .13) = 2(1 - .5517) = .8966$. There is no evidence to conclude that the process mean is not 224 mm.

6.32 (a) $H_0 : \mu = 9.5$, $H_a : \mu \neq 9.5$. **(b)** $z = (\bar{x} - \mu_0)/(\sigma/\sqrt{n}) = (9.57 - 9.5)/(.4/\sqrt{180}) = 2.35$. $P = 2P(Z > 2.35) = 2(1 - .9906) = .0188$. There is evidence to conclude that the mean differs from 9.5. **(c)** $\bar{x} \pm z^*\sigma/\sqrt{n} = 9.57 \pm 1.960(.4/\sqrt{180}) = (9.51, 9.63)$.

6.33 (a) $H_0 : \mu = 32$, $H_a : \mu > 32$. **(b)** $P = P(Z > 1.86) = (1 - .9686) = .0314$. There is evidence to conclude that the mean score of all third graders in this district is higher than the national mean.

6.34 (a) $z = (\bar{x} - \mu_0)/(\sigma/\sqrt{n}) = (.4365 - .5)/(.2887/\sqrt{100}) = -2.20$. **(b)** For a 5% test we use $z^* = 1.96$. Since $-2.20 < -1.96$ the result is significant at the 5% level. **(c)** For a 1% test we use $z^* = 2.576$. Since $-2.20 > -2.576$ the result is not significant at the 1% level.

6.35 (a) For a 5% test we use $z^* = 1.645$. Since $2.42 > 1.645$ the result is significant at the 5% level. **(b)** For a 1% test we use $z^* = 2.326$. Since $2.42 > 2.326$ the result is significant at the 1% level.

6.36 (a) For a 5% test we use $z^* = 1.960$. Since $-1.37 > -1.960$ the result is not significant at the 5% level. **(b)** Since the result is not significant at the 5% level, it is not significant at the 1% level.

6.37 If a significance test is significant at the 1% level the probability of observing a result as or more extreme than the result actually observed is less that 1%. Therefore, this probability is also less that 5%.

6.38 From Exercise 6.33, $z = -2.20$. From Table D for $p = .02$, $z^* = 2.054$ and for $p = .01$, $z^* = 2.326$. From this we can conclude that $P < .02$.

6.39 From Exercise 6.34, $z = 2.42$. From Table D for $p = .01$, $z^* = 2.326$ and for $p = .005$, $z^* = 2.576$. From this we can conclude that $P < .01$.

6.40 From Exercise 6.35, $z = -1.37$. From Table D for $p = .10$, $z^* = 1.282$ and for $p = .05$, $z^* = 1.645$. From Table A $P(Z < -1.37) = .0853$. Since this is a two-sided alternative, we double these numbers to get the P-value. Therefore, from Table D we conclude $.10 < P < .20$ and from Table A $P = 2(.0853) = .1706$.

6.41 (a) The 90% confidence interval is $(99.86, 108.41)$. **(b)** The hypotheses are $H_0 : \mu = 105$, $H_a : \mu \neq 105$. Since the 90% confidence interval includes 105, we do not reject H_0 at the 10% level.

6.42 (a) The 95% confidence interval from Exercise 6.5 is $(58, 62)$. **(b)** Since the 95% confidence interval includes 61.5, we do not reject $H_0 : \mu = 61.5$ versus the two–sided alternative at the 5% level. **(c)** Since 63 is not in the interval, we would reject this hypothesis.

6.43 (a) The 95% confidence interval is 4.2 ± 2.3 or $(1.9, 6.5)$. Because 7 is not in this interval, we reject $H_0 : \mu = 7$ in favor of $H_a : \mu \neq 7$. **(b)** Because 5 is in the confidence interval, we would not reject $H_0 : \mu = 5$ versus the two-sided alternative.

6.44 (a) $H_0 : p = .5$, $H_a : p > .5$. **(b)** X is $B(5, .5)$. **(c)** $P(X \geq 4) = P(X = 4) + P(X = 5) = 5(.5)^4(.5) + .5^5 = .1876$.

6.45 b.

CHAPTER 6

6.46 Statistical significance does not tell us whether an effect is strong or important.

6.47 (a) $z = (478-475)/(100/\sqrt{100}) = .30$. $P(Z > .30) = 1-.6179 = .3821$.
(b) $z = (478-475)/(100/\sqrt{1000}) = .9487$. $P(Z > .9487) = 1 - .8189 = .1711$. **(c)** $z = (478-475)/(100/\sqrt{10000}) = 3.00$. $P(Z > 3.00) = 1-.9987 = .0013$.

6.48 (a) $\bar{x} \pm z^*\sigma/\sqrt{n} = 478 \pm 2.576(100/\sqrt{100}) = (452.24, 503.76)$. **(b)** $\bar{x} \pm z^*\sigma/\sqrt{n} = 478 \pm 2.576(100/\sqrt{1000}) = (469.85, 486.15)$. **(c)** $\bar{x} \pm z^*\sigma/\sqrt{n} = 478 \pm 2.576(100/\sqrt{10000}) = (475.42, 480.58)$.

6.49 (a) $z = (491.4 - 475)/(100/\sqrt{100}) = 1.64$. $P(Z > 1.64) = 1 - .9495 = .0505$. The result is not significant at the 5% level. **(b)** $z = (491.5 - 475)/(100/\sqrt{100}) = 1.65$. $P(Z > 1.65) = 1 - .9505 = .0495$. The result is significant at the 5% level.

6.50 The station's conclusion is not justified. Voluntary response would bias the result.

6.51 (a) No, you would expect five results to be significant at the .01 level by chance alone. **(b)** They should be further tested.

6.52 (a) X is $B(77, .05)$. **(b)** $P(X \geq 2) = 1 - (P(X = 0) + P(X = 1)) = 1 - (.95^{77} + 77(.05).95^{76}) = 1 - (.0193 + .0781) = .9026$.

6.53 We reject H_0 if $(\bar{x} - 450)/(100/\sqrt{500}) \geq 2.326$ or $x \geq 460.4$. The power is $P(Z \geq (460.4 - 460)/(100/\sqrt{500})) = 1 - .5359 = .4641$. The test is not very sensitive to detect an increase of ten points.

6.54 (a) We reject H_0 if $(\bar{x} - 300)/(3/\sqrt{6}) \leq -1.645$ or $\bar{x} \leq 297.985$. The power is $P(Z \leq (297.985 - 299)/(3/\sqrt{6})) = .2033$. **(b)** The power is $P(Z \leq (297.985 - 295)/(3/\sqrt{6})) = .9927$. **(c)** The power against $\mu = 290$ will be higher than the value for part (b) because this alternative is farther away from the null hypothesized value of 300 than the alternative 295.

6.55 (a) We reject H_0 if $(\bar{x} - 300)/(3/\sqrt{25}) \leq -1.645$ or $\bar{x} \leq 299.013$.

The power is $P(Z \leq (299.013 - 299)/(3/\sqrt{25})) = .5080$. **(b)** We reject H_0 if $(\bar{x} - 300)/(3/\sqrt{100}) \leq -1.645$ or $\bar{x} \leq 299.5065$. The power is $P(Z \leq (299.5065 - 299)/(3/\sqrt{100})) = .9545$.

6.56 (a) We reject H_0 if $(\bar{x} - 128)/(15/\sqrt{72}) \leq -1.960$ or if $(\bar{x} - 128)/(15/\sqrt{72}) \geq 1.960$. This corresponds to $\bar{x} \leq 124.54$ or $\bar{x} \geq 131.46$. The power is $P(Z \leq (124.54 - 134)/(15/\sqrt{72})) + P(Z \geq (131.46 - 134)/(15/\sqrt{72})) = P(Z \leq -5.35) + P(Z \geq -1.44) = 0 + (1 - .0749) = .9251$. **(b)** The power is $P(Z \leq (124.54 - 122)/(15/\sqrt{72})) + P(Z \geq (131.46 - 122)/(15/\sqrt{72})) = P(Z \leq 1.44) + P(Z \geq 5.35) = .9251 + 0 = .9251$. The test has good power to detect a mean that differs from 128 by 6. **(c)** The power would be higher whenever the mean differs from 128 by more than 6. For example, when $\mu = 136$ the power is .9949.

6.57 (a) $P(\bar{x} > 0 | \mu = 0) = P(Z > (0 - 0)/(1/\sqrt{9})) = .50$. **(b)** $P(\bar{x} \leq 0 | \mu = .3) = P(Z \leq (0 - 3)/(1/\sqrt{9})) = .1841$. **(c)** $P(\bar{x} \leq 0 | \mu = 1) = P(Z \leq (0 - 1)/(1/\sqrt{9})) = .0013$.

6.58 The probability of a Type I error is $P(\bar{x} \leq 297.985 | \mu = 300) = .05$. The probability of a Type II error is $P(\bar{x} \geq 297.985 | \mu = 295) = 1 - .9927 = .0073$.

6.59 The probability of a Type I error is $P(\bar{x} \geq 460.4 | \mu = 450) = .01$. The probability of a Type II error is $P(\bar{x} \leq 460.4 | \mu = 460) = 1 - .4641 = .5359$.

6.60 (a) Using the p_0 distribution, we calculate the probability of a Type I error as $P(X = 0) + P(X = 1) + P(X = 2) + P(X = 3) + P(X = 5) = .1 + .1 + .1 + .1 + .1 = .5$. Using the p_1 distribution, we calculate the probability of a Type II error as $P(X = 4) + P(X = 6) = .2 + .1 = .3$.

6.61 (a) Patient does not need medical attention, patient needs medical attention. Referring a patient who does not need medical attention; clearing a patient who needs medical attention. **(b)** The second error probability should be made small. Failing to provide medical attention to a patient needing it could have very serious consequences.

6.62 (a) The probability the lot shipped is good is .1. The probability the lot shipped is bad is .9. The probability the lot is accepted given that it is bad is .08. The probability the lot is rejected given that it is bad is .92. The probability the lot is accepted given that it is good is .95. The probability the lot is rejected given that it is good is .05. **(b)** $P(accept) = P(accept \cap bad) + P(accept \cap good) = P(bad)P(accept|bad) + P(good)P(accept|good) = .9(.08) + .1(.95) = .1670$. **(c)** $P(bad|accepted) = P(bad \cap accepted)/P(accepted) = .0720/.1670 = .4311$.

6.63 The 90% confidence interval is $\bar{x} \pm z^*\sigma/\sqrt{n} = 5.37 \pm (1.645)(.9/\sqrt{6}) = (4.76, 5.98)$.

6.64 $P(\bar{x} \geq 5.37) = P(Z \geq (5.37 - 4.8)/(.9/\sqrt{6}) = P(Z \geq 1.54) = 1 - .9382 = .0618$. The data do not provide evidence to conclude that the mean phosphorus level exceeds 4.8.

6.65 (a)

```
2 | 034
2 |
3 | 01124
3 | 6
4 | 3
```

(b) The 95% confidence interval is $\bar{x} \pm z^*\sigma/\sqrt{n} = 30.4 \pm (1.96)(7/\sqrt{10}) = (26.06, 34.74)$. **(c)** The hypotheses are $H_0 : \mu = 25$ and $H_a : \mu > 25$. The test statistic is $z = (\bar{x} - \mu_0)/(\sigma/\sqrt{n}) = (30.4 - 25)/(7/\sqrt{10}) = 2.44$. The P-value is $P(Z \geq 2.44) = .0073$. We are convinced that the mean odor threshold for beginning students is higher than the published threshold.

6.66 (a) Narrower. If the probability that the interval covers the true value is less, then the interval is smaller. **(b)** The 99% confidence interval is given as $30853 \pm 397 = (30456, 31250)$. Since 32000 is not in this interval, we would reject the null hypothesis at the 1% level.

6.67 (a) $\bar{x} \pm z^*\sigma/\sqrt{n} = 145 \pm (1.645)(8/\sqrt{15}) = (141.6, 148.4)$. **(b)** $H_0 : \mu = 140$. $H_a : \mu > 140$. $z = (145 - 140)/(8/\sqrt{15}) = 2.42$. $P(Z > 2.42) = 1 - .9922 = .0078$. There is strong evidence that the mean cellulose content is greater than 140. **(c)** The data are an SRS from a normal population with known standard deviation.

6.68 $\bar{x} \pm z^*\sigma/\sqrt{n} = 12.9 \pm (1.960)(1.6/\sqrt{26}) = (12.285, 13.515)$. For this method we assume that we have an SRS from a normal distribution with known standard deviation.

6.69 (a) The width of the confidence interval is $2z^*\sigma/\sqrt{n}$. As n increases the width decreases because the $\sqrt{n}$ is in the denominator. **(b)** The P-value decreases. Roughly speaking, the z statistic $(\bar{x} - \mu_0)/(\sigma/\sqrt{n})$ will increase. **(c)** The power will increase.

6.70 $H_0 : p = 18/38$, $H_a : p \neq 18/38$.

6.71 The statement is false and misleading. The .05 is a probability statement about the test statistic, not a statement about the null hypothesis.

6.72 Yes. The chance variation in the statement is calculated under the assumption that the null hypothesis is true.

6.73 (a) Under the assumption that the two populations have the same proportions, the chance of seeing a difference as extreme as that observed is less than .01. **(b)** The probability that the interval will cover the true difference in population proportions is .95. **(c)** No. The participants in the program were volunteers.

6.74 (a) We drew 25 SRSs of size 100 from a normal distribution with mean 460 and standard deviation 100. There are a total of 2500 observations, too many to be listed. **(b)** The margin of error is $z^*\sigma/\sqrt{n} = 1.96(100)/\sqrt{100} = 19.6$. **(c)** There are 25 intervals. The first in our simulation is (455.77, 494.97). Here the margin of error is $(494.97 - 455.77)/2 = 19.6$. The intervals will vary but the margin of error will always be the same. **(d)** In our simulation all but one of the intervals contained the true mean. If we repeated the simulation we would not expect exactly the same number of intervals to contain μ. In a very large number of samples, 95% of the intervals would contain μ.

6.75 (a) The P-values in our simulation ranged from .003 to .493. **(b)** For the first sample the mean is 455.60, $z = -.44$, and $P = .32999$. From Table A, the entry corresponding to $z = -.44$ is .3300. **(c)** Of the 25 tests in our simulation, one rejected H_0. In a very large number of samples 5% would falsely reject the null hypothesis.

6.76 (a) There are 25 SRSs of 100 observations each giving a total of 2500 observations, too many to be listed. **(b)** Of the 25 tests in our simulation, 11 had P-values .05 or less. **(c)** We reject H_0 if $(\bar{x}-460)/(100/\sqrt{100}) \leq -1.960$ or if $(\bar{x} - 460)/(100/\sqrt{100}) \geq 1.960$. This corresponds to $\bar{x} \leq 440.4$ or $\bar{x} \geq 479.6$. The power is $P(Z \leq (440.4-480)/(100/\sqrt{100}))+P(Z \geq (479.6-480)/(100/\sqrt{100})) = P(Z \leq -3.96) + P(Z \geq -0.04) = 0 + (1 - .4840) = .5160$. In a very large number of samples from a population with mean 480, we would expect 51.6% of the samples to reject the null hypothesis.

4.7 CHAPTER 7

7.1 (a) $df = 9$, $t^* = 2.262$. (b) $df = 19$, $t^* = 2.861$. (c) $df = 6$, $t^* = 1.440$.

7.2 (a) $df = 14$. (b) From Table E, $t^* = 1.761$ corresponds to $p = .05$ and $t^* = 2.145$ corresponds to $p = .025$. (c) $.025 \leq P \leq .05$. (d) It is significant at the 5% level but not at the 1% level.

7.3 (a) $df = 24$. (b) From Table E, $t^* = 1.059$ corresponds to $p = .15$ and $t^* = 1.318$ corresponds to $p = .10$. (c) $.20 \leq P \leq .30$. (d) It is not significant at the 5% level or the 1% level.

7.4 (a) $df = 9$. (b) From Table E, $t^* = 1.833$ corresponds to $p = .05$ and $t^* = 2.262$ corresponds to $p = .025$. Therefore, $.025 \leq P \leq .05$.

7.5 $s/\sqrt{n} = 37/\sqrt{4} = 18.5$.

7.6 (a) $\bar{x} \pm t^* s/\sqrt{n} = 24 \pm 2.00(11/\sqrt{75}) = (21.46, 26.54)$. (b) The sample mean of 75 observations will have a distribution that is nearly normal even if the population is not normal.

7.7 (a) $H_0 : \mu = 0$, $H_a : \mu > 0$. $t = (\bar{x} - \mu_0)/(s/\sqrt{n}) = (332 - 0)/(108/\sqrt{200}) = 43.47$. $t^* = 2.364$ for $df = 100$. There is significant evidence at the 1% level that the mean amount charged increases under the no-fee offer.
(b) $\bar{x} \pm t^* s/\sqrt{n} = 332 \pm 2.626(108/\sqrt{200}) = (312.0, 352.1)$. (Using the exact t^* from a computer, the interval is $(312.139, 351.861)$.) (c) When the sample size is large the t procedure can be used for skewed distributions. (d) Randomly assign the customers to two groups. One group will receive the no-fee offer, and the other will not receive the offer.

7.8 (a) $\bar{x} = 5.37$. $s_{\bar{x}} = s/\sqrt{n} = .665/\sqrt{6} = .272$. (b) $\bar{x} \pm t^* s/\sqrt{n} = 5.37 \pm 2.015(.665/\sqrt{6}) = (4.82, 5.92)$.

7.9 (a) $\bar{x} = 1.75$. $s_{\bar{x}} = s/\sqrt{n} = .129/\sqrt{4} = .065$. (b) $\bar{x} \pm t^* s/\sqrt{n} = 1.75 \pm 2.353(.129/\sqrt{4}) = (1.60, 1.90)$.

7.10 $H_0 : \mu = 4.8$. $H_a : \mu > 4.8$. $t = (\bar{x} - \mu_0)/(s/\sqrt{n}) = (5.37 - 4.8)/(.665/\sqrt{6}) = 2.08$. From Table E, $.025 \leq P \leq .05$. There is significant evidence at the 5% level to conclude that the patient's phosphate level is higher than normal.

7.11 $H_0 : \mu = 1.3$. $H_a : \mu > 1.3$. $t = (\bar{x} - \mu_0)/(s/\sqrt{n}) = (1.75 - 1.3)/(.129/\sqrt{4}) = 6.97$. From Table E, $.0025 \leq P \leq .005$. There is significant evidence at the .5% level to conclude that the mean absolutely refractory period for these rats is larger than the mean for unpoisoned rats.

7.12 (a) $\bar{x} \pm t^*s/\sqrt{n} = 114.9 \pm 2.056(9.3/\sqrt{27}) = (111.22,\ 118.58)$. (b) We assume that we have a simple random sample from a normal population. The assumption that the observations are independent is most important.

7.13 (a) $\bar{x} \pm t^*s/\sqrt{n} = 1.67 \pm 2.120(.25/\sqrt{17}) = (1.54,\ 1.80)$. (b) An SRS is selected from a population with a normal distribution. The SRS is most important.

7.14 (a)

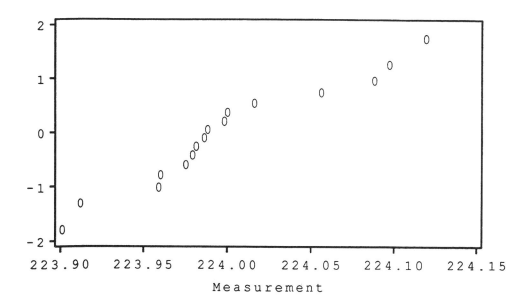

(b) $H_0 : \mu = 224$. $H_a : \mu \neq 224$. $t = (\bar{x} - \mu_0)/(s/\sqrt{n}) = (224.002 -$

$224)/(.0618/\sqrt{16}) = .129$. From Table E, $P(T \leq .691) = .25$. Since this is a two-sided test, we conclude that $P > .50$. The exact probability is $P = .90$.

7.15 (a)

```
 9 | 2
 9 | 578
10 | 024
10 | 55
11 | 1
11 | 9
12 | 2
```

(b) $H_0 : \mu = 105$. $H_a : \mu \neq 105$. $t = (\bar{x} - \mu_0)/(s/\sqrt{n}) = (104.133 - 105)/(9.397/\sqrt{12}) = -.319$. From Table E, $P(T \leq .697) = .25$. Since this is a two-sided test, we conclude that $P > .50$. The exact probability is $P = .76$.

7.16 The estimated mean response is $\bar{x} = 22.125$. We choose to give a margin of error corresponding to a 95% confidence interval. $t^*s/\sqrt{n} = 3.182(2.090/\sqrt{4}) = 3.33$. If we repeated this experiment many times, 95% of the time the true mean would be within the interval defined by the observed mean plus or minus the margin of error.

7.17 (a) Standard error of the mean. **(b)** $s/\sqrt{n} = s/\sqrt{3} = .01$. This implies $s = .0173$. **(c)** $\bar{x} \pm t^*s/\sqrt{n} = .84 \pm 2.920(.0173/\sqrt{3}) = (.811, .869)$.

7.18 (a) Flip a coin. If it lands heads, do the right thread and then the left thread. If it lands tails, do the opposite. **(b)** Let X be the difference between the time for the right thread and the time for the left thread. We assume that we have a random sample of X's from a normal population with a mean μ. $H_0 : \mu = 0$. $H_a : \mu < 0$. $t = (\bar{x} - \mu_0)/(s/\sqrt{n}) = (-13.32 - 0)/(22.936/\sqrt{25}) = -2.90$. From Table E, $.0025 \leq P \leq .005$. The exact value is $P = .0039$.

7.19 $\bar{x} \pm t^*s/\sqrt{n} = -13.32 \pm 1.711(22.936/\sqrt{25}) = (-21.17, -5.47)$. No, the average time saved is rather small. The mean time for right hand threads as a percent of the mean time for left hand threads is 88.66%.

7.20 (a) Let X be the difference between the post-test and the pre-test. We assume that we have a random sample of X's from a normal population with a mean μ. $H_0 : \mu = 0$. $H_a : \mu > 0$. **(b)** The data are skewed to the left. We are not greatly concerned about a small amount of skewness with a sample size of 20.

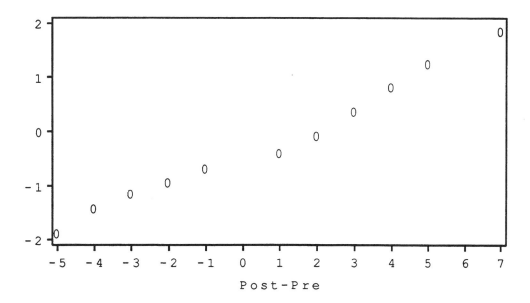

(c) $t = (\bar{x} - \mu_0)/(s/\sqrt{n}) = (1.45 - 0)/(3.2032/\sqrt{20}) = 2.024$. From Table E, $.025 \leq P \leq .050$. The exact probability is $P = .029$. We reject at the 5% level but not at the 1% level. **(d)** $\bar{x} \pm t^* s/\sqrt{n} = 1.45 \pm 1.729(3.2032/\sqrt{20}) = (.2115, 2.6885)$.

7.21 (a) Toss a coin to decide which test is given first for each person. **(b)** $t = (\bar{x} - \mu_0)/(s/\sqrt{n}) = (.2519 - 0)/(.2767/\sqrt{22}) = 4.27$. This is a two-sided test. Therefore, from Table E, $P < .001$. The exact probability is $P = .0003$. The two tests do not have the same means. **(c)** $\bar{x} \pm t^* s/\sqrt{n} = .2519 \pm 2.080(.2767/\sqrt{22}) = (.129, .375)$.

7.22 Let μ be the difference in the means. $H_0 : \mu = 0$. $H_a : \mu > 0$. $t = (\bar{x} - \mu_0)/(s/\sqrt{n}) = (1.32 - 0)/(2.35/\sqrt{9}) = 1.69$. From Table E, $.05 \leq P \leq .10$. The exact probability is $P = .065$. The observed difference in means is marginally significant. **(b)** $\bar{x} \pm t^* s/\sqrt{n} = 1.32 \pm 1.860(2.35/\sqrt{9}) =$

$(-.14, 2.78)$.

7.23 $H_0 : \mu_A = \mu_B$, $H_a : \mu_A > \mu_B$. $t = (\bar{x} - \mu_0)/(s/\sqrt{n}) = (.34 - 0)/(.83/\sqrt{10}) = 1.30$. From Table E, $.10 \leq P \leq .15$. The exact probability is $P = .113$. There is no convincing evidence that variety A has the higher yield.

7.24 (a) Two independent samples. **(b)** Paired comparison. **(c)** Single sample. **(d)** Paired comparison.

7.25 The results are based on a census and there is no sample.

7.26 (a) $df = 49$. From Table E we use $t^* = 2.403$, the entry for $df = 50$. **(b)** We reject H_0 if $t \geq 2.403$. First, $t = (\bar{x} - \mu_0)/(s/\sqrt{n}) = (\bar{x} - 0)(108/\sqrt{50})$. Therefore, the rejection region in terms of $\bar{x}$ is $\bar{x} \geq 36.70$. **(c)** $P(\bar{x} \geq 36.70 | \mu = 100) = P(Z \geq (36.70 - 100)/(108/\sqrt{50})) = P(Z \geq -4.144)$. From Table E, we conclude that the power is greater than $(1 - .0005) = .9995$. The actual power is .999983. This power is very high. There is no need to use more than 50 customers in the study.

7.27 (a) $df = 9$. From Table E we use $t^* = 1.833$. We reject H_0 if $t \geq 1.833$. First, $t = (\bar{x} - \mu_0)/(s/\sqrt{n}) = (\bar{x} - 0)(.83/\sqrt{10})$. Therefore, the rejection region in terms of $\bar{x}$ is $\bar{x} \geq .4811$. The power is $P(\bar{x} \geq .4811 | \mu = .5) = P(Z \geq (.4811 - .5)/(.83/\sqrt{10})) = P(Z \geq -.0720) = 1 - .4721 = .5279$. **(b)** $df = 24$. From Table E we use $t^* = 1.711$. We reject H_0 if $t \geq 1.711$. First, $t = (\bar{x} - \mu_0)/(s/\sqrt{n}) = (\bar{x} - 0)(.83/\sqrt{25})$. Therefore, the rejection region in terms of $\bar{x}$ is $\bar{x} \geq .284$. The power is $P(\bar{x} \geq .284 | \mu = .5) = P(Z \geq (.284 - .5)/(.83/\sqrt{25})) = P(Z \geq -1.30) = 1 - .0968 = .9032$.

7.28 (a) $df = 21$. The critical values for a two-sided test with $\alpha = .05$ are $t^* = 2.080$ and $t^* = -2.080$. **(b)** We reject H_0 if $t \geq 2.080$ or if $t \leq -2.080$. First, $t = (\bar{x} - \mu_0)/(s/\sqrt{n}) = (\bar{x} - 0)(.3/\sqrt{22})$. Therefore, the rejection region in terms of $\bar{x}$ is $\bar{x} \geq .133$ or $\bar{x} \leq -.133$. **(c)** The power is $P(\bar{x} \geq .133 | \mu = .2) + P(\bar{x} \leq -.133 | \mu = .2) = P(Z \geq (.133 - .2)/(.3/\sqrt{22})) + P(Z \leq (-.133 - .2)/(.3/\sqrt{22})) = 1 - .1469 + 0 = .8531$.

7.29 (a) $H_0 : \eta = 0$, $H_a : \eta < 0$, $H_0 : p = .5$, $H_a : p > .5$. **(b)**

First, we delete the case where both values are 105. Of the remaining 24 cases, 19 have right less than left. The P-value is $P(X \geq 19) = P(X = 19) + P(X = 20) + \ldots + P(X = 24) = .0033$. The normal approximation is $P((X - np)/\sqrt{np(1-p)} \geq (19-12)/\sqrt{24(.5)(.5)}) = 1 - .9979 = .0021$. With the continuity correction the value is .0040. The right hand thread can be completed faster than the left hand thread.

7.30 (a) $H_0 : \eta = 0$, $H_a : \eta > 0$, $H_0 : p = .5$, $H_a : p > .5$. (b) Of the 20 cases, 14 improved. The P-value is $P(X \geq 14) = P(X = 14) + P(X = 15) + \ldots + P(X = 20) = .0577$. These values can be found in Table C. The normal approximation is $P((X-np)/\sqrt{np(1-p)} \geq (14-10)/\sqrt{20(.5)(.5)}) = 1 - .9633 = .0367$. With the continuity correction the value is .0582. The evidence is marginal that attending the institute improves listening skills. Comparing this result with that obtained in Exercise 7.20, we see that the sign test is less powerful than the t test.

7.31 The test cannot be done because the number of people with positive differences is not given.

7.32 $H_0 : \eta = 0$, $H_a : \eta > 0$, $H_0 : p = .5$, $H_a : p > .5$. Let X represent the number of locations where Variety A had the higher yield. Under H_0, X is $B(10, .5)$. The P-value is $P(X \geq 6) = P(X = 6) + P(X = 7) + \ldots + P(X = 10) = .3770$. These values can be found in Table C. There is no evidence in these data to support the conclusion that Variety A has a higher median yield.

7.33 (a) From Table E, we use the $t^* = 2.000$ for $df = 60$. $\bar{x} \pm t^*s/\sqrt{n} = 141.847 \pm 2.000(109.21/\sqrt{72}) = (116.11, 167.59)$. The exact interval using $df = 71$ and $t^* = 1.99394$ is $(116.18, 167.51)$. (b) (c) $\bar{x} \pm t^*s/\sqrt{n} = 2.072 \pm 2.000(.243/\sqrt{72}) = (2.015, 2.129)$. The interval rounded in this way agrees with the interval obtained using the exact degrees of freedom.

7.34 The logs are 2.4771, 2.5105, 2.5705, 2.5705, and 2.6474. The confidence interval is $\bar{x} \pm t^*s/\sqrt{n} = 2.555 \pm 2.132(.0653/\sqrt{5}) = (2.49, 2.62)$.

7.35 $H_0 : \mu_1 = \mu_2$, $H_a : \mu_1 > \mu_2$. $t = (\bar{x}_1 - \bar{x}_2)/\sqrt{(s_1^2/n_1) + (s_2^2/n_2)} =$

$(3.47 - 1.36)/\sqrt{(1.21^2/13) + (.52^2/14)} = 5.81$. The minimum of $n_1 - 1$ and $n_2 - 1$ is 12. Using Table E, $t^* = 2.681$ for a one-sided test at the 1% level with $df = 12$. There is significant evidence to conclude that the mean number of larvae per stem is reduced by malathion.

7.36 (a) $H_0 : \mu_1 = \mu_2$, $H_a : \mu_1 \neq \mu_2$. $t = (\bar{x}_1 - \bar{x}_2)/\sqrt{(s_1^2/n_1) + (s_2^2/n_2)} = (4.64 - 6.43)/\sqrt{(.69^2/14) + (.43^2/14)} = -8.24$. The minimum of $n_1 - 1$ and $n_2 - 1$ is 13. Using Table E, $t^* = 2.160$ for a two-sided test at the 5% level with $df = 13$, and $t^* = 3.012$ for a two-sided test at the 1% level with $df = 13$. **(b)** We cannot generalize to the population of all middle-aged men because the subjects in this study were volunteers.

7.37 (a) $(\bar{x}_1 - \bar{x}_2) \pm t^*\sqrt{(s_1^2/n_1) + (s_2^2/n_2)} = (52 - 49) \pm 2.000\sqrt{(13^2/75) + (11^2/53)} = (-1.3, 7.3)$. The minimum of $n_1 - 1$ and $n_2 - 1$ is 52. Using Table E, $t^* = 2.009$ for a 95% confidence interval with $df = 50$. An exact interval using $t^* = 2.00665$ for $df = 52$ is $(-1.274, 7.274)$. **(b)** We are 95% confident that the true change in sales is covered by our interval. Since the interval includes 0 and negative values, we cannot be certain that the sales have increased.

7.38 (a) $H_0 : \mu_A = \mu_B$, $H_a : \mu_A \neq \mu_B$. $t = (\bar{x}_A - \bar{x}_B)/\sqrt{(s_A^2/n_A) + (s_B^2/n_B)} = (1987 - 2056)/\sqrt{(392^2/150) + (413^2/150)} = -1.48$. The minimum of $n_A - 1$ and $n_B - 1$ is 149. Using Table E with $df = 100$, we conclude $.10 \leq P \leq .20$. The exact value is $P = .13989$. These data contain no evidence to conclude that the two proposals are different. **(b)** For the large samples in this problem the two-sample t procedure is robust against skewness.

7.39 (a) $H_0 : \mu_1 = \mu_2$, $H_a : \mu_1 > \mu_2$. $t = (\bar{x}_1 - \bar{x}_2)/\sqrt{(s_1^2/n_1) + (s_2^2/n_2)} = (13.3 - 12.4)/\sqrt{(1.7^2/23) + (1.8^2/19)} = 1.65$. df=18, $P = .057$ (from Table E P-value is between .05 and .10). The difference between the hemoglobin levels is not significant at the .05 level. **(b)** $(\bar{x}_1 - \bar{x}_2) \pm t^*\sqrt{(s_1^2/n_1) + (s_2^2/n_2)} = (13.3 - 12.4) \pm 2.101\sqrt{(1.7^2/23) + (1.8^2/19)} = (-.24, 2.04)$. The minimum of $n_1 - 1$ and $n_2 - 1$ is 18. **(c)** The two means are calculated from independent

SRSs from normal populations. The standard deviations are not assumed to be equal.

7.40 $H_0 : \mu_1 = \mu_2$, $H_a : \mu_1 > \mu_2$. Where μ_1 is the population mean for the control group. $t = (\bar{x}_1 - \bar{x}_2)/\sqrt{(s_1^2/n_1) + (s_2^2/n_2)} = (70.3 - 65.2)/\sqrt{(8.3^2/30) + (7.8^2/30)} = 2.45$. Using Table E for $df = 29$ we conclude $.01 \le P \le .02$. The exact value is $P = .0102205$. The result is significant at the 5% level but not at the 1% level. $(\bar{x}_1 - \bar{x}_2) \pm t^*\sqrt{(s_1^2/n_1) + (s_2^2/n_2)} = (70.3 - 65.2) \pm 2.756\sqrt{(8.3^2/30) + (7.8^2/30)} = (-.63, 10.83)$.

7.41 (a) $H_0 : \mu_1 = \mu_2$, $H_a : \mu_1 > \mu_2$. Where μ_1 is the population mean for the skilled rowers. **(b)** From the output we find $t = 3.1583$. The two–sided P–value is given as .0104 for the unequal variance t statistic and the degrees of freedom are 9.8. For our one–sided alternative we check that the means differ in the direction specified by the alternative hypothesis and then divide by 2 giving $P = .0052$. We conclude that the skilled rowers have a higher mean angular knee velocity. Note that we obtain the same conclusion if we use the equal variance procedure. **(c)** For the confidence interval we use the conservative (giving a slightly larger interval) degrees of freedom, minimum of $8 - 1 = 7$ and $10 - 1 = 9$ to obtain the value of t^*. The interval is $(\bar{x}_1 - \bar{x}_2) \pm t^*\sqrt{(s_1^2/n_1) + (s_2^2/n_2)} = (4.18 - 3.01) \pm 1.895\sqrt{(.48^2/10) + (.96^2/8)} = (.47, 1.87)$.

7.42 The hypotheses are $H_0 : \mu_1 = \mu_2$, $H_a : \mu_1 \ne \mu_2$. Where μ_1 is the population mean for the skilled rowers. From the output we find $t = .5143$. The degrees of freedom for the unequal variance test are 9.8 and the $P = .6184$. There is no evidence to conclude that there is a difference in mean weights. Note that we obtain the same conclusion if we use the equal variance procedure.

7.43 $(\bar{x}_1 - \bar{x}_2) \pm t^*\sqrt{(s_1^2/n_1) + (s_2^2/n_2)} = (416 - 386) \pm 2.576\sqrt{(87^2/19883) + (74^2/19937)} = (27.91, 32.09)$. Using Table E, $t^* = 2.576$ for a 99% confidence interval with $df = \infty$.

7.44 (a) The data for the males are slightly skewed to the right. The value

200 is rather large for the females.

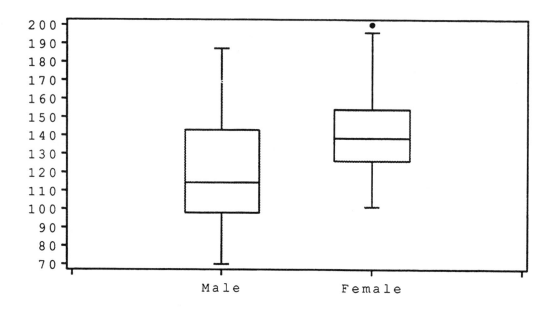

Here is a back-to-back stemplot for the two sexes. Males are on the left.

```
    50 |  7 |
     8 |  8 |
    21 |  9 |
   984 | 10 | 139
  5543 | 11 | 5
     6 | 12 | 669
     2 | 13 | 77
    60 | 14 | 08
     1 | 15 | 244
     9 | 16 | 55
       | 17 | 8
    70 | 18 |
       | 19 |
       | 20 | 0
```

(b) $H_0 : \mu_1 = \mu_2$, $H_a : \mu_1 > \mu_2$. Where μ_1 is the population mean for the women. $t = (\bar{x}_1 - \bar{x}_2)/\sqrt{(s_1^2/n_1) + (s_2^2/n_2)} = (141.06 -$

$121.25)/\sqrt{(26.4363^2/18) + (32.8519^2/20)} = 2.06$. From Table E we conclude $.025 \leq P \leq .050$. The exact value is $P = .028$. There is evidence to conclude that the mean SSHA of the women is higher. (c) A 90% confidence interval for the female minus male difference in means is $(\bar{x}_1 - \bar{x}_2) \pm t^*\sqrt{(s_1^2/n_1) + (s_2^2/n_2)} = (141.06 - 121.25) \pm 1.74\sqrt{(26.4363^2/18) + (32.8519^2/20)} = (3.05, 36.57)$. The minimum of $n_1 - 1$ and $n_2 - 1$ is 17.

7.45 (a) There are no outliers. The distributions are approximately symmetric.

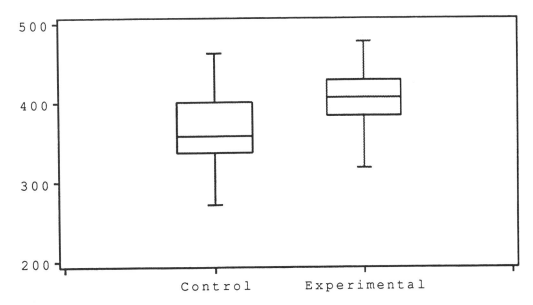

(b) $H_0 : \mu_1 = \mu_2$, $H_a : \mu_1 < \mu_2$. Where μ_1 is the population mean for the control group and μ_2 is the population mean for the experimental group. $t = (\bar{x}_2 - \bar{x}_1)/\sqrt{(s_2^2/n_2) + (s_1^2/n_1)} = (366.30 - 402.95)/\sqrt{(50.81^2/20) + (42.73^2/20)} = -2.46$. From Table E we conclude $.01 \leq P \leq .02$. The exact value is $P = .012$. The difference is significant at the 10% and the 5% levels but not at the 1% level. (c) A 95% confidence interval for the mean extra weight gain in chicks fed high-lysine corn is $(\bar{x}_1 - \bar{x}_2) \pm t^*\sqrt{(s_1^2/n_1) + (s_2^2/n_2)} = (402.95 - 366.30) \pm 2.093\sqrt{(42.73^2/20) + (50.81^2/20)} = (5.58, 67.72)$.

The minimum of $n_1 - 1$ and $n_2 - 1$ is 19.

7.46 (a) $H_0 : \mu_1 = \mu_2$, $H_a : \mu_1 > \mu_2$. Where μ_1 is the population mean for the calcium group and μ_2 is the population mean for the placebo group. $t = (\bar{x}_1 - \bar{x}_2)/\sqrt{(s_1^2/n_1) + (s_2^2/n_2)} = (5.000 - (-.273))/\sqrt{(8.74^2/10) + (5.90^2/11)} = 1.61$. From Table E we conclude $.05 \leq P \leq .10$. The exact value is $P = .0716$. **(b)** A 90% confidence interval for the difference in population means is $(\bar{x}_1 - \bar{x}_2) \pm t^*\sqrt{(s_1^2/n_1) + (s_2^2/n_2)} = (5.000 - (-.273)) \pm 1.833\sqrt{(8.74^2/10) + (5.90^2/11)} = (-.747, 11.293)$. This interval is wider.

7.47 (a) A 95% confidence interval for the difference in the mean calorie content between beef and poultry hot dogs. $(\bar{x}_1 - \bar{x}_2) \pm t^*\sqrt{(s_1^2/n_1) + (s_2^2/n_2)} = (156.75 - 122.47) \pm 2.120\sqrt{(22.51^2/20) + (25.48^2/17)} = (17.38, 51.18)$. **(b)** Yes. The confidence interval does not include 0. **(c)** The two means are calculated from independent SRSs from normal populations. The standard deviations are not assumed to be equal. No clear deviation from the normal assumption is evident.

7.48 (a) The standard error of the sample mean VOT for the adults is $s_{\bar{x}_2} = s_2/\sqrt{n_2} = 50.74/\sqrt{20} = 11.3458$. The standard error of the difference is $\sqrt{(s_1^2/n_1) + (s_2^2/n_2)} = \sqrt{(33.89^2/10) + (50.74^2/20)} = 15.61$. **(b)** $H_0 : \mu_1 = \mu_2$, $H_a : \mu_1 \neq \mu_2$. $t = (\bar{x}_1 - \bar{x}_2)/\sqrt{(s_1^2/n_1) + (s_2^2/n_2)} = (-3.67 - (-23.17))/15.61 = 1.25$ From Table E we conclude $.20 \leq P \leq .30$. The exact value is $P = .243$. Based on the data we have no evidence to conclude that the mean scores for children and adults are different. **(c)** A 95% confidence interval for the difference between the scores of children and adults is $(\bar{x}_1 - \bar{x}_2) \pm t^*\sqrt{(s_1^2/n_1) + (s_2^2/n_2)} = (-3.67 - (-23.17)) \pm 2.262(15.61) = (-15.8, 54.8)$. We knew the interval would contain zero because the null hypothesis was not reject at the 5% level.

7.49 If there are no differences between the children and adults, approximately 5% of the tests would be significant.

7.50 (a) A 90% confidence interval for the difference in population means is $(\bar{x}_1 - \bar{x}_2) \pm t^*\sqrt{(s_1^2/n_1) + (s_2^2/n_2)} = (1884.52 - 1360.39) \pm 1.646\sqrt{(1368.37^2/675) + (1037.46^2/621)} = (414, 635)$. **(b)** The large sample sizes justify the use of the t procedures. **(c)** Yes, there should be no systematic bias in an alphabetical list. **(d)** Since the students who were not employed were dropped, the results do not apply to all undergraduates. We would also like to know the percentage of students who did not respond to the questionnaire.

7.51 $t = (\bar{x}_1 - \bar{x}_2)/\sqrt{(s_1^2/n_1) + (s_2^2/n_2)} = (17.6 - 9.5)/\sqrt{(6.34^2/6) + (1.95^2/6)} = 2.99$. $df = ((6.34^2/6) + (1.95^2/6))^2/(((6.34^2/6)^2/5) + ((1.95^2/6)^2/5)) = 5.93$.

7.52 $s_p^2 = ((n_1 - 1)s_1^2 + (n_2 - 1)s_2^2)/(n_1 + n_2 - 2) = ((132)5.05^2 + (161)5.44^2)/293 = 27.7505$. $s_p = \sqrt{27.75} = 5.2679$. $t = (\bar{x}_1 - \bar{x}_2)/s_p\sqrt{(1/n_1) + (1/n_2)} = (25.34 - 24.94)/5.2679\sqrt{(1/133) + (1/162)} = .6489$. $df = n_1 + n_2 - 2 = 293$. Using Table E with $df = 100$, we conclude $P > .5$. The exact value is $P = .516893$. The results are essentially the same as those obtained in Example 8.9.

7.53 (a) $H_0: \mu_1 = \mu_2$, $H_a: \mu_1 > \mu_2$. $s_p^2 = ((n_1 - 1)s_1^2 + (n_2 - 1)s_2^2)/(n_1 + n_2 - 2) = ((22)1.7^2 + (18)1.8^2)/40 = 3.0475$. $s_p = \sqrt{3.0475} = 1.7457$. $t = (\bar{x}_1 - \bar{x}_2)/s_p\sqrt{(1/n_1) + (1/n_2)} = (13.3 - 12.4)/1.7457\sqrt{(1/23) + (1/19)} = 1.66$. $df = n_1 + n_2 - 2 = 40$. $P = .052$ (from Table E P-value is between .05 and .10). The difference between the hemoglobin levels is not significant at the .05 level. **(b)** A 95% confidence interval for the difference in population means is $(\bar{x}_1 - \bar{x}_2) \pm t^*s_p\sqrt{(1/n_1) + (1/n_2)} = (13.3 - 12.4) \pm 2.021(.5412)\sqrt{(1/23) + (1/19)} = (-.19, 1.99)$. The results are essentially the same.

7.54 $s_p^2 = ((n_1 - 1)s_1^2 + (n_2 - 1)s_2^2)/(n_1 + n_2 - 2) = ((9)6.1^2 + (7)9.04^2)/16 = 56.6838$. $s_p = \sqrt{56.6838} = 7.52887$. $t = (\bar{x}_1 - \bar{x}_2)/s_p\sqrt{(1/n_1) + (1/n_2)} = (70.37 - 68.45)/7.52887\sqrt{(1/10) + (1/8)} = .5376$. $df = n_1 + n_2 - 2 = 16$. From Table E we conclude $P > .5$. The exact value is $P = .5982$. The results are essentially the same as those obtained in Exercise 7.42.

7.55 (a) $s_p^2 = ((n_1 - 1)s_1^2 + (n_2 - 1)s_2^2)/(n_1 + n_2 - 2) = ((9)33.89^2 + (19)50.74^2)/28 = 2116.19$. $s_p = \sqrt{2116.19} = 17.8165$. $t = (\bar{x}_1 - \bar{x}_2)/s_p\sqrt{(1/n_1) + (1/n_2)} = (-3.67 - (-23.17))/17.8165\sqrt{(1/10) + (1/20)} = 1.09$. $df = n_1 + n_2 - 2 = 28$. From Table E we conclude $.20 \leq P \leq .30$. The exact value is $P = .28$. **(b)** A 95% confidence interval for the difference in population means is $(\bar{x}_1 - \bar{x}_2) \pm t^* s_p\sqrt{(1/n_1) + (1/n_2)} = (-3.67 - (-23.17)) \pm 2.048(17.8165)\sqrt{(1/10) + (1/20)} = (-17.00, 56.00)$. **(c)** The results are very similar.

7.56 (a) The df are the minimum of $n_1 - 1$ and $n_2 - 1$ or 99. Using Table E with $df = 100$ we obtain $t^* = 2.364$. **(b)** Using $(\bar{x}_1 - \bar{x}_2)/\sqrt{(8^2/100) + (8^2/100)} \geq 2.364$, we reject H_0 when $(\bar{x}_1 - \bar{x}_2) \geq 2.6746$. **(c)** The power is $P((\bar{x}_1 - \bar{x}_2) \geq 2.6746|(\mu_1 - \mu_2) = 5) = P(Z \geq (2.6746 - 5)/\sqrt{(8^2/100) + (8^2/100)} = 1 - .0197 = .9803$.

7.57 (a) $H_0 : \mu_1 = \mu_2$, $H_a : \mu_1 \neq \mu_2$. $t = (\bar{x}_1 - \bar{x}_2)/\sqrt{(s_1^2/n_1) + (s_2^2/n_2)}$. **(b)** For an $\alpha = .05$ test we reject H_0 when t is greater than 1.984, or less than -1.984. (We use the entrry corresponding to 100 degrees of freedom from Table E.) **(c)** Using $(\bar{x}_1 - \bar{x}_2)/\sqrt{(30^2/100) + (50^2/300)} \geq 1.984$, we reject H_0 when $(\bar{x}_1 - \bar{x}_2) \geq 8.2601$. or when $(\bar{x}_1 - \bar{x}_2) \leq -8.2601$. The power is $P((\bar{x}_1 - \bar{x}_2) \geq 8.2601|(\mu_1 - \mu_2) = 10) + P((\bar{x}_1 - \bar{x}_2) \leq -8.2601|(\mu_1 - \mu_2) = 10) = P(Z \geq (8.2601 - 10)/\sqrt{(30^2/100) + (50^2/300)} + P(Z \leq (-8.2601 - 10)/\sqrt{(30^2/100) + (50^2/300)} = 1 - .0084 + 0 = .9916$.

7.58 (a) $H_0 : \mu_A - \mu_B = 0$, $H_a : \mu_A - \mu_B \neq 0$. $t = (\bar{x}_1 - \bar{x}_2)/\sqrt{(s_1^2/n_1) + (s_2^2/n_2)}$. **(b)** The $\alpha = .05$ critical values are 1.962 and -1.962 for $df = 1000$. **(c)** Using $(\bar{x}_1 - \bar{x}_2)/\sqrt{(400^2/350) + (400^2/350)} \geq 1.962$, we reject H_0 when $(\bar{x}_1 - \bar{x}_2) \geq 59.3253$. or when $(\bar{x}_1 - \bar{x}_2) \leq -59.3253$. The power is $P((\bar{x}_1 - \bar{x}_2) \geq 59.3253|(\mu_1 - \mu_2) = 100) + P((\bar{x}_1 - \bar{x}_2) \leq -59.3253|(\mu_1 - \mu_2) = 100) = P(Z \geq (59.3253 - 100)/\sqrt{(400^2/350) + (400^2/350)} + P(Z \leq (-59.3253 - 100)/\sqrt{(400^2/350) + (400^2/350)} = 1 - .0885 + 0 = .9115$.

7.59 (a) The degrees of freedom are 9 and 7. The 5% upper critical value is 3.68. This corresponds to an $\alpha = .1$ test. (b) The result is not significant at the 10% or the 5% level.

7.60 (a) The degrees of freedom are 20 and 15. The 2.5% upper critical value is 2.76. Since 2.78 is greater than this value, the result is significant at the 5% level. The 1% upper critical value is 3.37. Since 2.78 is less than this value, the result is not significant at the 2% level. Therefore, the result is not significant at the 1% level. (b) From Table F we conclude $.02 \leq P \leq .05$.

7.61 $F = 10.58$. $df = (5, 5)$. From Table F, the P-value is between .02 and .05. The hypothesis of equal population standard deviations is rejected at the 5% level but not at the 1% level.

7.62 (a) $H_0 : \sigma_1 = \sigma_2$, $H_a : \sigma_1 \neq \sigma_2$. (b) $F = .96^2/.48^2 = 3.98$. $df = (7, 9)$. From Table F, the P-value is between .05 and .10.

7.63 (a) $H_0 : \sigma_1 = \sigma_2$, $H_a : \sigma_1 \neq \sigma_2$. (b) $F = 9.04^2/6.10^2 = 2.20$. $df = (7, 9)$. The P-value is greater than .2.

7.64 $F = 50.74^2/33.89^2 = 2.24$. $df = (19, 9)$. Using Table F with $df = (20, 9)$ we conclude $P > .2$. Using a computer, the exact value is $P = .215$. There is no evidence to conclude that the standard deviations are different.

7.65 (a) $H_0 : \sigma_1 = \sigma_2$, $H_a : \sigma_1 < \sigma_2$, where group 1 is the females. (b) $F = s_2^2/s_1^2 = 32.85^2/26.44^2 = 1.54$. (c) Since there is no entry for 19 df in the numerator, we use the entries for $F(20, 17)$. $P > .1$. The exact value is $P = .186$. There is no clear evidence that the men are more variable than the women.

7.66 $F = 87^2/74^2 = 1.38$. $df = (415, 385)$. Using Table F for $df = (120, 200)$ we conclude $.02 \leq P \leq .05$. Using Table F for $df = (1000, 1000)$ we conclude $P < .002$. The exact value is $P = .0013$. Since we have very large sample sizes, we are able to detect moderate differences with high power.

7.67 (a) .045. (b) .075.

7.68 (a) .035. (b) .055. (c) .085.

7.69 (a) The change is weight is the variable of interest. Therefore, the pairs are the before and after measurements. (b) The mean weight change for these subjects was a weight loss. We are not given information on the amount lost. (c) From Table E, we can conclude $P < .0005$.

7.70 (a) We use t^* corresponding to 1000 degrees of freedom from Table E. Note that we can use the standard errors of the means given in the output to simplify our calculations slightly. The 99% confidence interval is $\mu_1 - \mu_2 \pm t^* \sqrt{s_{\bar{x}_1}^2 + s_{\bar{x}_2}^2} = 7368 - 6595 \pm 2.581\sqrt{7.8309^2 + 6.6132^2} = 1043 \pm 26 = (1017, 1069)$. (b) The t procedures can be used in the presence of strong skewness because we have very large sample sizes.

7.71 $H_0 : \mu_1 = \mu_2$, $H_a : \mu_1 > \mu_2$, where μ_1 is the population mean for the control group and μ_2 is the population mean for the nitrite group. $t = (\bar{x}_1 - \bar{x}_2)/\sqrt{(s_1^2/n_1) + (s_2^2/n_2)} = (8112 - 7880)/\sqrt{(1115^2/30) + (1250^2/30)} = .7586$. Using $df = 29$ from Table E we conclude $.20 \leq P \leq .25$. The exact value is $P = .2271$. There is no evidence to conclude that nitrites decrease amino acid uptake. The same conclusion is reached if equation (7.4) is used or if the pooled two-sample procedure is used.

7.72 $H_0 : \mu_1 = \mu_2$, $H_a : \mu_1 < \mu_2$, where μ_1 is the population mean for the students and μ_2 is the population mean for the workers. $t = (\bar{x}_1 - \bar{x}_2)/\sqrt{(s_1^2/n_1) + (s_2^2/n_2)} = (35.12 - 37.32)/\sqrt{(4.31^2/750) + (3.83^2/412)} = -8.95$. Using Table E with $df = 1000$ we conclude $P < .0005$. (b) With large samples the t procedure can be used. (c) For a normal distribution 95% of the values will be between $\mu \pm 1.96\sigma$. Although this distribution is skewed, we will base our interval on this approximation. $\bar{x}_2 \pm 1.96\sigma_2 = 37.32 \pm 1.96(3.83) = (29.81, 44.83)$. (d) The distribution for the fifteenth minute is more variable and more symmetric.

7.73 (a) No clear skewness or outliers are evident.

```
4 | 9
5 | 1
5 | 4
```

6 | 03344
6 | 5

(b) $\bar{x} \pm t^*s/\sqrt{n} = 59.59 \pm 2.306(6.255)/\sqrt{9} = (54.78, 64.40)$.

7.74 (a) Standard error of the mean. To calculate s we use $s_{\bar{x}} = s/\sqrt{n}$; for example, $44 = s/\sqrt{98}$ implies $s = 436$. Drivers: 2821, 436; .24, .59. Conductors: 2844, 437; .39, 1.00. (b) We let population 1 be the conductors and population 2 be the drivers. $t = (\bar{x}_1 - \bar{x}_2)/\sqrt{(s_1^2/n_1) + (s_2^2/n_2)} = (2844 - 2821)/\sqrt{(437.30^2/83) + (435.58^2/98)} = .3532$. $df = 82$. The difference is not significant at the 5% level. (c) $t = (\bar{x}_1 - \bar{x}_2)/\sqrt{(s_1^2/n_1) + (s_2^2/n_2)} = (.39 - .24)/\sqrt{(1.00^2/83) + (.59^2/98)} = 1.20$. $df = 82$. $P = .12$. (d) $\bar{x}_1 \pm t^*s_1/\sqrt{n} = .39 \pm 1.664(1.00)/\sqrt{83} = (.21, .57)$. (e) $\bar{x}_1 - \bar{x}_2 \pm t^*\sqrt{(s_1^2/n_1) + (s_2^2/n_2)} = .39 - .24 \pm 1.292\sqrt{(1.00^2/83) + (.59^2/98)} = (-.012, .312)$. This is the confidence interval for the alcohol consumption of conductors minus the alcohol consumption of drivers. An interval for drivers minus conductors is $(-.312, .012)$.

7.75 The pooled two-sample t test is justified because the two sample standard deviations are quite close ($s_1 = 436$, $s_2 = 437$). $s_p = \sqrt{(n_1 - 1)(s_1^2) + (n_2 - 1)(s_2^2)} = \sqrt{(97)(435.58^2) + (82)(437.30^2)} = 436.37$. Therefore, $t = (\bar{x}_1 - \bar{x}_2)/s_p\sqrt{(1/n_1) + (1/n_2)} = (2844 - 2821)/436.37\sqrt{(1/98) + (1/83)} = .35$. The degrees of freedom are 179. The exact P-value is .36. The results are essentially the same as those obtained in the previous exercise.

7.76 (a) No. With large samples skewness does not invalidate the test. (b) Yes. This test is very sensitive to the normality assumption.

7.77 (a) $H_0 : \sigma_1 = \sigma_2$, $H_a : \sigma_1 \neq \sigma_2$. $F = s_2^2/s_1^2 = 5.44^2/5.05^2 = 1.16$. The degrees of freedom are 161 and 132. The values in Table F for degrees of freedom close to these are all larger than the calculated value of 1.16. Therefore, we have no evidence to reject the hypothesis that the two standard deviations are equal. The exact P-value is .375. (b) If the distributions are not normal, the test for equality of standard deviations is not recommended.

7.78 No. This is a census. The mean population can be computed with no error.

7.79 (a) There are no obvious outliers. There is no strong skewness.
(b) $\bar{x} \pm t^*s/\sqrt{n} = 907.75 \pm 2.131(8.481)/\sqrt{16} = (903, 912)$. (c) $t = (\bar{x} - \mu_0)/(s/\sqrt{n}) = (907.75 - 910)/(8.48/\sqrt{16}) = -1.06$. Using Table E with $df = 15$ we conclude $.30 \leq P \leq .40$. The exact value is $P = .3054$.

7.80 (a) $t = (\bar{x} - \mu_0)/(s/\sqrt{n}) = (83 - 86)/(10/\sqrt{40}) = -1.90$. The degrees of freedom are 39. The result is not significant at the 1% level. (b) The two methods should be tested in a comparative experiment.

7.81 (a) $H_0 : \mu_1 = \mu_2$, $H_a : \mu_1 > \mu_2$, where μ_1 is the population mean for the pets and μ_2 is the population mean for the clinic dogs. $t = (\bar{x}_1 - \bar{x}_2)/\sqrt{(s_1^2/n_1) + (s_2^2/n_2)} = (193 - 174)/\sqrt{(68^2/26) + (44^2/23)} = 1.17$. Using Table E with $df = 22$ we conclude $.10 \leq P \leq .15$. The exact value is $P = .13$. There is no evidence to conclude that the mean cholesterol level is different in pets and clinic dogs. (b) $\bar{x}_1 - \bar{x}_2 \pm t^*\sqrt{(s_1^2/n_1) + (s_2^2/n_2)} = 193 - 174 \pm 2.074\sqrt{(68^2/26) + (44^2/23)} = (-14.6, 52.6)$. (c) $\bar{x}_1 \pm t^*s_1/\sqrt{n_1} = 193 \pm 2.060(68)/\sqrt{26} = (166, 220)$. (d) We assume that the samples are SRS's from the respective populations and that the distributions are normal. The chief threat to the validity of the study is that the data are not from random samples.

7.82 (a) $H_0 : \mu_1 = \mu_2$, $H_a : \mu_1 > \mu_2$. (b) A paired comparison is appropriate. (c) For this problem we analyze the city value minus the rural value. There are two rather high outliers. The following stemplot was produced by SAS.

```
-0 | 32111100
 0 | 11111112222222
 0 | 57
 1 |
 1 | 58
```

(d) $t = (\bar{x} - \mu_0)/(s/\sqrt{n}) = (2.19 - 0)/(4.69/\sqrt{26}) = 2.38$. Using Table E with $df = 25$, we conclude that $.01 \leq P \leq .02$. The exact value is $P = .012$. There is evidence that the city particulate level exceeds the rural level. (e)

$\bar{x} \pm t^*s/\sqrt{n} = 2.19 \pm 1.708(4.69)/\sqrt{26} = (.619, 3.761)$.

7.83 $H_0 : \eta = 0$, $H_a : \eta > 0$, $H_0 : p = .5$, $H_a : p > .5$. Let X represent the number of days on which city levels are higher than the rural levels. Under H_0, X is $B(26, .5)$. The P-value is $P(X \geq 18) = P(X = 18) + P(X = 19) + \ldots + P(X = 26) = .03776$. These values can be calculated using the formula for the binomial probabilities or computer software. We conclude that the particulate levels are higher in the city than in the rural area.

7.84 $\bar{x} = 552.63$. Note that the sample size is 27 days. For 95% confidence the margin of error is $t^*s/\sqrt{n} = 2.056(152.98)/\sqrt{27} = 60.53$.

7.85 There is one clear outlier, 4.07, and another value which is rather low, 4.88. Excluding these two observations, the mean is 5.49. For 95% confidence, the margin of error is $t^*s/\sqrt{n} = 2.056(.193)/\sqrt{27} = .08$. By contrast, if we do not discard the two outliers, the mean is 5.42 and the margin of error is .13.

```
40 | 7
41 |
42 |
43 |
44 |
45 |
46 |
47 |
48 | 8
49 |
50 |
51 | 0
52 | 6799
53 | 04469
54 | 2467
55 | 03578
56 | 1235
57 | 59
58 | 56
```

7.86 For the difference between others and runners the 95% confidence intervals are Abdomen: $\bar{x}_1 - \bar{x}_2 \pm t^*\sqrt{(s_1^2/n_1) + (s_2^2/n_2)} = 20.6 - 7.1 \pm 2.093\sqrt{(9^2/95) + (1^2/20)} = (11.51, 15.49)$, Thigh: $\bar{x}_1 - \bar{x}_2 \pm t^*\sqrt{(s_1^2/n_1) + (s_2^2/n_2)} = 17.4 - 6.1 \pm 2.093\sqrt{(6.6^2/95) + (1.8^2/20)} = (9.65, 12.95)$.

7.87 (a) The data are approximately normal. There appears to be one high outlier. Here is the stemplot.

```
 4 | 0
 5 | 17
 6 | 0
 7 | 128
 8 | 6
 9 | 57
10 | 86
11 | 25667
12 | 02468
13 | 1122389
14 | 455779
15 | 1
16 |
17 | 389
18 | 037
19 |
20 | 1
21 |
22 | 07
23 |
24 |
25 |
26 |
27 |
28 |
29 | 4
```

Here is the normal quantile plot.

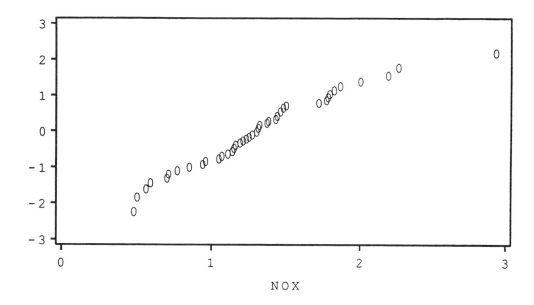

(b) The 99% confidence interval is (1.14, 1.52). This is generated by software using the exact value $t^* = 2.690$. (c) The test statistic is $t = 4.60$, with 45 degrees of freedom. The P-value is less than .0005. We are 99% confident that the mean NOX level is between 1.14 and 1.52. There is strong evidence that the mean is greater than 1.

7.88 (a) To compare the SATM averages for males and females using the pooled procedure, we find $t = 3.9969$ with 222 degrees of freedom. The P-value is .0001. For the separate sample procedure we find $t = 4.0124$, with 162.2 degrees of freedom. The P-value is .0001. The equality of standard deviations test statistic is $F = 1.03$ with 144 and 78 degrees of freedom. The P-value is .9114. It is appropriate to use the pooled procedure because the standard deviations are very close and the test indicates that the variances are not distinguishable. We conclude that males average higher scores than females on SATM. A 99% confidence interval for the male–female difference in means is (16.36, 77.13).

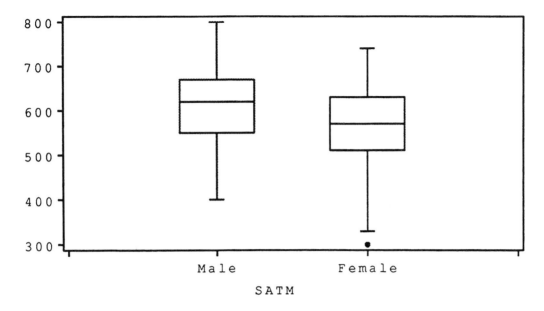

(b) To compare the SATV averages for males and females using the pooled procedure, we find $t = .9395$ with 222 degrees of freedom. The P-value is .3485. For the separate sample procedure we find $t = .9547$, with 167.9 degrees of freedom. The P-value is .3411. The equality of standard deviations test statistic is $F = 1.11$ with 144 and 78 degrees of freedom. The P-value is .6033. It is appropriate to use the pooled procedure because the standard deviations are very close and the test indicates that the variances are not distinguishable. We are unable to see any difference between males and females on SATM. A 99% confidence interval for the male–female difference in means is $(-21.48, 45.82)$.

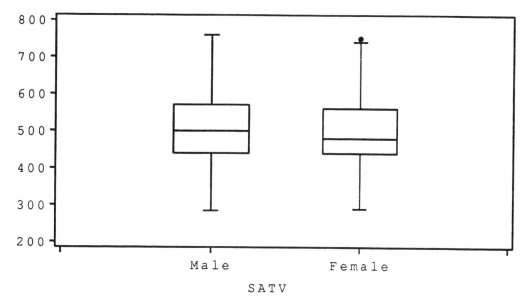

(c) Generalization of these results to other groups of students involves inference of a non–statistical nature. The more that the group resembles our sample, the more likely it is that the generalization will be approximately valid. Students entering the same university in different years are most similar, followed by computer science majors elsewhere. College students in general mey be very different from the students in our sample.

7.89 (a) The distribution is very discrete (with values $-.02, -.01, 0, .01, .02,$ and $.03$). There are no outliers and no skewness is evident. The mean is .0046 and the standard deviation is .0149. (b) The 95% confidence interval is $(.00037, .00883)$. (c) The test statistic is $t = 2.187$, with 49 degrees of freedom. The P-value is .0336. There is evidence to suggest that the second measure tends to be lower than the first measure.

7.90 (a) Th mean reading score with DRTA appears to be higher than for the Basal method. Pooling is justified. The pooled test results are $t = 2.8679$ with 42 degrees of freedom and $P = .0032$. We divided the two–sided result by 2 to obtain this P-value. The difference in the average scores is 5.68. (b) Th mean reading score with DRTA appears to be higher than for the Strat method. Pooling is justified. The pooled test results are $t = 1.8773$ with 42 degrees of freedom and $P = .0337$. We divided the two–sided result by 2 to obtain this P-value. The difference in the average scores is 3.23.

4.8 CHAPTER 8

8.1 (a) No. $np_0 = 10(.4) = 4$ and $n(1 - p_0) = 10(.6) = 6$. (b) Yes. $np_0 = 100(.6) = 60$ and $n(1 - p_0) = 100(.4) = 40$. (c) No. $np_0 = 1000(.996) = 996$ and $n(1 - p_0) = 1000(.004) = 4$. (d) Yes. $np_0 = 500(.3) = 150$ and $n(1 - p_0) = 500(.7) = 350$.

8.2 (a) No. $n\hat{p} = 30(.9) = 27$ and $n(1 - \hat{p}) = 30(.1) = 3$. (b) Yes. $n\hat{p} = 25(.5) = 12.5$ and $n(1 - \hat{p}) = 25(.5) = 12.5$. (c) No. $n\hat{p} = 100(.04) = 4$ and $n(1 - \hat{p}) = 100(.96) = 96$. (d) Yes. $n\hat{p} = 600(.6) = 360$ and $n(1 - \hat{p}) = 600(.4) = 240$.

8.3 The 95% confidence interval is $\hat{p} \pm z^*\sqrt{\hat{p}(1-\hat{p})/n} = 86/100 \pm 1.96\sqrt{.86(1-.86)/100} = .86 \pm .068 = (.792, .928)$.

8.4 The 99% confidence interval is $\hat{p} \pm z^*\sqrt{\hat{p}(1-\hat{p})/n} = 41/216 \pm 2.576\sqrt{(41/216)(1-41/216)/216} = .190 \pm .069 = (.121, .259)$.

8.5 $\hat{p} \pm z^*\sqrt{\hat{p}(1-\hat{p})/n} = .66 \pm 1.960\sqrt{.66(1-.66)/200} = (.595, .725)$.

8.6 (a) $\hat{p} \pm z^*\sqrt{\hat{p}(1-\hat{p})/n} = .42 \pm 2.576\sqrt{.42(1-.42)/1785} = (.39, .45)$. (b) $H_0 : p = .5$. $H_a : p < .5$. $z = (\hat{p} - p_0)/\sqrt{p_0(1-p_0)/n} = (.42 - .5)/\sqrt{.5(1-.5)/1785} = -6.78$. Using Table A we find that $P < .0002$. There is good evidence to conclude that less than half of the population attended church or synagogue. (c) Since $\alpha = 1 - .99 = .01$, the upper $\alpha/2$ critical value, z^*, is 2.576. The width, w, is twice the margin of error. $n = (2z^*/w)^2 p^*(1-p^*) = (2(2.576)/.02)^2 .42(1-.42) = 16164.75$. Therefore, a sample of size 16165 is needed.

8.7 Since $\alpha = 1 - .95 = .05$, the upper $\alpha/2$ critical value, z^*, is 1.960. The width, w, is twice the margin of error. $n = (2z^*/w)^2 p^*(1-p^*) = (2(1.960)/.06)^2 .44(1-.44) = 1051.74$. Therefore, a sample of size 1052 is needed.

8.8 $\hat{p} \pm z^*\sqrt{\hat{p}(1-\hat{p})/n} = .1733 \pm 1.960\sqrt{.1733(1-.1733)/75} = (.09, .26)$.

8.9 (a) We want to test $H_0 : p = .5$ versus the alternative $H_a : p > .5$. The proportion of home games won is $532/(532 + 438) = .5485$. The test statistic is $z = (\hat{p} - p_0)/\sqrt{p_0(1 - p_0)/n} = (.5485 - .5)/\sqrt{.5(1 - .5)/970} = 3.01$. Using Table A we find that $P = 1 - .9987 = .0013$. There is convincing evidence in favor of a home team advantage. **(b)** The 95% confidence interval is $\hat{p} \pm z^*\sqrt{\hat{p}(1 - \hat{p})/n} = .5485 \pm 1.960\sqrt{.5485(1 - .5485)/970} = .5485 \pm .0314(.5171\ .580)$. There is a slight advantage in favor of the home team.

8.10 $H_0 : p = .48$. $H_a : p \neq .48$. $z = (\hat{p} - p_0)/\sqrt{p_0(1 - p_0)/n} = (.44 - .48)/\sqrt{.48(1 - .48)/500} = -1.79$. Using Table A we find that $P = 2(.0367) = .0734$. We have no evidence to conclude that the telephone survey technique has a probability of selecting a household with no children that is different from the proportion reported in the 1980 Census.

8.11 (a) $H_0 : p = .5$. $H_a : p \neq .5$. $z = (\hat{p} - p_0)/\sqrt{p_0(1 - p_0)/n} = (.5005 - .5)/\sqrt{.5(1 - .5)/24000} = .15$. Using Table A we find that $P = 2(1 - .5596) = .88$. For $\alpha = .01$, z^* is 2.576. Therefore, Pearson's coin is indistinguishable from a fair coin on the basis of 24,000 tosses. **(b)** $\hat{p} \pm z^*\sqrt{\hat{p}(1 - \hat{p})/n} = .5005 \pm 2.576\sqrt{.5005(1 - .5005)/24000} = (.492, .509)$.

8.12 (a) The hypotheses are $H_0 : p = .5$. $H_a : p \neq .5$. The test statistic is $z = (\hat{p} - p_0)/\sqrt{p_0(1 - p_0)/n} = (.5067 - .5)/\sqrt{.5(1 - .5)/10000} = 1.34$. Using Table A we find that $P = 2(1 - .9099) = .18$. For $\alpha = .05$, z^* is 1.96. Therefore, Kerrich's coin is indistinguishable from a fair coin on the basis of 10,000 tosses. **(b)** The 95% confidence interval is $\hat{p} \pm z^*\sqrt{\hat{p}(1 - \hat{p})/n} = .5067 \pm 1.96\sqrt{.5067(1 - .5067)/10000} = (.4969, .5165)$.

8.13 (a) The hypotheses are $H_0 : p = .5$. $H_a : p < .5$. The test statistic is $z = (\hat{p} - p_0)/\sqrt{p_0(1 - p_0)/n} = (.38 - .5)/\sqrt{.5(1 - .5)/50} = -1.70$. Using Table A we find that $P = .0446$. We reject the null hypothesis that instant coffee tastes just as good as fresh brewed. The difference is significant at the 5% level because $P < .05$. The difference in proportions is .12. This result may be practically important to a company that makes instant coffee and is seeking to improve the taste to make it more like fresh brewed. **(b)** A 90% confidence interval for p is $\hat{p} \pm z^* \sqrt{\hat{p}(1-\hat{p})/n} = .38 \pm 1.645\sqrt{.38(1-.38)/50} =$ (.267, .493).

8.14 (a) $H_0: p = .4$. $H_a: p > .4$. **(b)** $z = (\hat{p} - p_0)/\sqrt{p_0(1-p_0)/n} = (.625 - .4)/\sqrt{.4(1-.4)/40} = 2.90$. **(c)** For $\alpha = .05$, $z^* = 1.645$. Therefore, we reject H_0 for $\alpha = .05$. Using Table A we find that $P = 1 - .9981 = .0019$. We conclude that his free throw percentage has increased. **(d)** The 90% confidence interval is $\hat{p} \pm z^* \sqrt{\hat{p}(1-\hat{p})/n} = .625 \pm 1.645\sqrt{.625(1-.625)/40} = (.50, .75)$. Because the interval does not include .4, and we have rejected the null hypothesis with $P = .0019$ we are convinced that LeRoy has improved his free throw shooting. **(e)** We assume that the shots represent independent trials with the same probability of success. For the normal approximation used in the test we need $np_0 = 40(.4) = 16 > 10$ and $n(1-p_0) = 40(.6) = 24 > 10$. Similarly, for the confidence interval we need $n\hat{p} = 40(.625) = 25 > 10$ and $n(1-\hat{p}) = 40(.375) = 15 > 10$. All of these conditions are satisfied.

8.15 Since $\alpha = 1 - .95 = .05$, the upper $\alpha/2$ critical value, z^*, is 1.96. $n = (z^*/m)^2 p^*(1-p^*) = (1.96/.05)^2 .3(1-.3) = 322.683$. Therefore, a sample of size 323 is needed.

8.16 Since $\alpha = 1 - .99 = .01$, the upper $\alpha/2$ critical value, z^*, is 2.576. $n = (z^*/m)^2 p^*(1-p^*) = (2.576/.025)^2 .5(1-.5) = 2654.31$. Therefore, a sample of size 2655 is needed.

8.17 Since $\alpha = 1 - .90 = .1$, the upper $\alpha/2$ critical value, z^*, is 1.645. $n = (z^*/m)^2 p^*(1-p^*) = (1.645/.04)^2 .7(1-.7) = 355.17$. Therefore, a sample of size 356 is needed. $m = z^* \sqrt{\hat{p}(1-\hat{p})/n} = (1.645)\sqrt{.5(1-.5)/356} = .044$.

8.18 Since $\alpha = 1 - .99 = .01$, the upper $\alpha/2$ critical value, z^*, is 2.576. $n = (2z^*/w)^2 p^*(1-p^*) = (2(2.576)/.03)^2 .2(1-.2) = 4718.15$. Therefore, a sample of size 4718 is needed. $w = 2z^* \sqrt{\hat{p}(1-\hat{p})/n} = 2(2.576)\sqrt{.1(1-.1)/4718} = .02$.

8.19 (a) Using the formula, $m = z^* \sqrt{\hat{p}(1-\hat{p})/n}$ we calculate the margins of error as .059, .079, .090, .096, .098, .096, .090, .079, .059. **(b)** No. np is too small.

8.20 The margins of error for a sample of size 500 are .027, .035, .040, .043, .044, .043, .040, .035, .027. The normal approximation could be used for $p = .04$ with this sample size because now $np = 500(.04) = 20$. The larger sample size is better because the margins of error are smaller.

8.21 (a) The proportions are $51/81 = .6296$ for home games and $44/81 = .5432$ for away games.
(b)

$$\begin{aligned} s_D &= \sqrt{(\hat{p}_1(1-\hat{p}_1)/n_1) + (\hat{p}_2(1-\hat{p}_2)/n_2)} \\ &= \sqrt{(.6296(1-.6296)/81) + (.5432(1-.5432)/81)} \\ &= .0771 \end{aligned}$$

(c)

$$\begin{aligned} \hat{p}_1 - \hat{p}_2 \pm z^* s_D &= .6296 - .5432 \pm 1.645(.0771) \\ &= (-.040, .213) \end{aligned}$$

Since this confidence interval includes the value zero, we conclude that the Twins were not more likely to win at home. The difference in the proportions is indistinguishable from chance variation.

8.22 (a) Tippecanoe: $263/515 = .511$, Benton: $260/637 = .408$.
(b)

$$\begin{aligned} s_D &= \sqrt{(\hat{p}_1(1-\hat{p}_1)/n_1) + (\hat{p}_2(1-\hat{p}_2)/n_2)} \\ &= \sqrt{(.511(1-.511)/515) + (.408(1-.408)/637)} \\ &= .0294 \end{aligned}$$

(c)

$$\begin{aligned} \hat{p}_1 - \hat{p}_2 \pm z^* s_D &= .511 - .408 \pm 2.576(.0294) \\ &= (.03, .18) \end{aligned}$$

8.23 (a) Combining all games played, the Twins proportion of wins is $\hat{p} = (X_1 + X_2)/(n_1 + n_2) = (51 + 44)/(81 + 81) = .586$.
(b)

$$\begin{aligned} s_p &= \sqrt{\hat{p}(1-\hat{p})((1/n_1) + (1/n_2))} \\ &= \sqrt{.586(1-.586)((1/81) + (1/81))} \\ &= .077 \end{aligned}$$

(c) $H_0 : p_1 = p_2$, $H_a : p_1 > p_2$.
(d)

$$\begin{aligned} z &= (\hat{p}_1 - \hat{p}_2)/s_p \\ &= (.6296 - .5432)/.077 \\ &= 1.12 \end{aligned}$$

From Table A we find $P = 1 - .8686 = .1314$. There is not sufficient evidence to conclude that it is easier to win at home.

8.24 (a) $\hat{p} = (X_1 + X_2)/(n_1 + n_2) = (263 + 260)/(515 + 637) = .454$.
(b)

$$\begin{aligned} s_p &= \sqrt{\hat{p}(1-\hat{p})((1/n_1) + (1/n_2))} \\ &= \sqrt{.454(1-.454)((1/515) + (1/637))} \\ &= .0295 \end{aligned}$$

(c) $H_0 : p_1 = p_2$, $H_a : p_1 \neq p_2$.
(d)

$$\begin{aligned} z &= (\hat{p}_1 - \hat{p}_2)/s_p \\ &= (.5107 - .4082)/.0295 \\ &= 3.47 \end{aligned}$$

From Table A we find $P = 2(1 - .9997) = .0006$. The exact value is $P = .00051$. We conclude that farmers in Tippecanoe county are more in favor of the corn checkoff program than farmers in Benton county.

8.25 We let p_1 denote the proportion of birth defects when the well was in use and p_2 denote the proportion of birth defects when the well was shut off. We want to test $H_0 : p_1 = p_2$, versus $H_a : p_1 > p_2$.
The sample proportions are $\hat{p}_1 = 16/414 = .0386$ and $\hat{p}_2 = 3/228 = .0132$. Combining all of the data, we have $\hat{p} = (16 + 3)/(414 + 228) = .0296$. The pooled standard error is

$$\begin{aligned} s_p &= \sqrt{\hat{p}(1-\hat{p})((1/n_1) + (1/n_2))} \\ &= \sqrt{.0296(1-.0296)((1/414) + (1/228))} \\ &= .0140 \end{aligned}$$

The test statistic is

$$\begin{aligned} z &= (\hat{p}_1 - \hat{p}_2)/s_p \\ &= (.0386 - .0132)/.0140 \\ &= 1.81 \end{aligned}$$

From Table A we find $P = 1 - .9649 = .0354$. We conclude that the birth defect rate decreased when the well was shut off. Our analysis assumes that the data come from two independent binomial populations. Since the pooled $\hat{p} = .03$, the values of n_1 and n_2 in this problem are sufficiently large for the normal approximation to be reasonably accurate. The assumptions are not fully met in this problem. Some of the same families may have had births during both periods, so independence is probably violated. The real question is whether or not they are sufficiently reasonable to lead one to the conclusion that the well was a cause of birth defects. Here we are concerned about other changes that may have happened over time. This problem could serve as the basis for some interesting class discussion.

8.26 The data table is

	Criminal	Not Criminal	All
Normal	381	3715	4096
Abnormal	8	20	28
Sum	389	3735	4124

The table of row percents is

	Criminal	Not Criminal
Normal	9.30	90.70
Abnormal	28.57	71.43
Sum	9.43	90.57

We test $H_0 : p_{ij} = r_i c_j$ versus the alternative $H_a : p_{ij} \neq r_i c_j$ for some i and j. Using formula (8.2) we calculate $X^2 = 12.09$ with one degree of freedom. From Table G we find $P = .001$. We conclude that these abnormalities are associated with higher criminality. This problem could be formulated as a comparison of two proportions with a one-sided alternative. In this case we would find $z = 3.45$ with $P = .0006$.

8.27 $\hat{p}_1 = .603$, $\hat{p}_2 = .591$, $\hat{p} = (X_1 + X_2)/(n_1 + n_2) = (161 + 136)/(267 + 230) = .598$.

$$\begin{aligned} s_p &= \sqrt{\hat{p}(1-\hat{p})((1/n_1) + (1/n_2))} \\ &= \sqrt{.598(1-.598)((1/267) + (1/230))} \\ &= .0441 \end{aligned}$$

$H_0 : p_1 = p_2$, $H_a : p_1 \neq p_2$.

$$\begin{aligned} z &= (\hat{p}_1 - \hat{p}_2)/s_p \\ &= (.603 - .591)/.0441 \\ &= .27 \end{aligned}$$

From Table A we find $P = 2(1-.6064) = .79$. The exact value is $P = .79099$. There is no evidence that the Protestants and Catholics differ on this issue.

8.28 $\hat{p}_1 = 104/267 = .3895$, $\hat{p}_2 = 75/230 = .3261$,

$$\begin{aligned} s_D &= \sqrt{(\hat{p}_1(1-\hat{p}_1)/n_1) + (\hat{p}_2(1-\hat{p}_2)/n_2)} \\ &= \sqrt{(.3895(1-.3895)/267) + (.3261(1-.3261)/230)} \\ &= .0430 \end{aligned}$$

$$\hat{p}_1 - \hat{p}_2 \pm z^* s_D = .3895 - .3261 \pm 1.960(.0430)$$
$$= (-.0208, .1476)$$

8.29 (a) $H_0 : p_1 = p_2$, $H_a : p_1 \neq p_2$. $\hat{p}_1 = 718/797 = .901$, $\hat{p}_2 = 593/732 = .810$, $\hat{p} = (X_1 + X_2)/(n_1 + n_2) = (718 + 593)/(797 + 732) = .857$.

$$s_p = \sqrt{\hat{p}(1-\hat{p})((1/n_1) + (1/n_2))}$$
$$= \sqrt{.857(1-.857)((1/797) + (1/732))}$$
$$= .0179$$

$$z = (\hat{p}_1 - \hat{p}_2)/s_p$$
$$= (.901 - .810)/.0179$$
$$= 5.07$$

Using Table A we find $P < .0004$. The exact value is $P = .0000004$.
(b)

$$s_D = \sqrt{(\hat{p}_1(1-\hat{p}_1)/n_1) + (\hat{p}_2(1-\hat{p}_2)/n_2)}$$
$$= \sqrt{(.901(1-.901)/797) + (.810(1-.810)/732)}$$
$$= .018$$

$$\hat{p}_1 - \hat{p}_2 \pm z^* s_D = .901 - .810 \pm 2.576(.018)$$
$$= (.045, .137)$$

8.30 (a) $H_0 : p_1 = p_2$, $H_a : p_1 \neq p_2$.
$\hat{p}_1 = 35/83 = .4217$. $\hat{p}_2 = 15/136 = .1103$. $\hat{p} = (X_1 + X_2)/(n_1 + n_2) = (35 + 15)/(83 + 136) = .2283$.

$$s_p = \sqrt{\hat{p}(1-\hat{p})((1/n_1) + (1/n_2))}$$
$$= \sqrt{.2283(1-.2283)((1/83) + (1/136))}$$
$$= .0585$$

$$z = (\hat{p}_1 - \hat{p}_2)/s_p$$
$$= (.4217 - .1103)/.0585$$
$$= 5.33$$

Using Table A we find $P < .0004$. The exact value is $P = .0000001$.
(b)
$$s_D = \sqrt{(\hat{p}_1(1-\hat{p}_1)/n_1) + (\hat{p}_2(1-\hat{p}_2)/n_2)}$$
$$= \sqrt{(.4217(1-.4217)/83) + (.1103(1-.1103)/136)}$$
$$= .0605$$

$$\hat{p}_1 - \hat{p}_2 \pm z^* s_D = .4217 - .1103 \pm 1.645(.0605)$$
$$= (.2119, .4109)$$

8.31 (a) $\hat{p}_1 = 63/78 = .808$. $\hat{p}_2 = 43/77 = .558$.
(b)
$$s_D = \sqrt{(\hat{p}_1(1-\hat{p}_1)/n_1) + (\hat{p}_2(1-\hat{p}_2)/n_2)}$$
$$= \sqrt{(.808(1-.808)/78) + (.558(1-.558)/77)}$$
$$= .072$$

$$\hat{p}_1 - \hat{p}_2 \pm z^* s_D = .808 - .558 \pm 1.960(.072)$$
$$= (.108, .391)$$

(c) $H_0: p_1 = p_2$, $H_a: p_1 > p_2$. $\hat{p} = (X_1 + X_2)/(n_1 + n_2) = (63+43)/78+77) = .684$.

$$s_p = \sqrt{\hat{p}(1-\hat{p})((1/n_1) + (1/n_2))}$$
$$= \sqrt{.684(1-.684)((1/78) + (1/77))}$$
$$= .075$$

$$z = (\hat{p}_1 - \hat{p}_2)/s_p$$
$$= (.808 - .558)/.075$$
$$= 3.34$$

CHAPTER 8

Using Table A $P = 1 - .9996 = .0004$. Use of aspirin increases the probability of a favorable outcome for patients with cerebral ischemia.

8.32 (a) Let treatment one be plasterboard and treatment two be glass. $\hat{p}_1 = 13/18 = .7222$. $\hat{p}_2 = 9/18 = .5000$.
(b)

$$\begin{aligned} s_D &= \sqrt{(\hat{p}_1(1-\hat{p}_1)/n_1) + (\hat{p}_2(1-\hat{p}_2)/n_2)} \\ &= \sqrt{(.7222(1-.7222)/18) + (.5000(1-.5000)/18)} \\ &= .1582 \end{aligned}$$

$$\begin{aligned} \hat{p}_1 - \hat{p}_2 \pm z^* s_D &= .7222 - .5000 \pm 1.645(.1582) \\ &= (-.0381, .4825) \end{aligned}$$

(c) $H_0 : p_1 = p_2$, $H_a : p_1 > p_2$. $\hat{p} = (X_1+X_2)/(n_1+n_2) = (13+9)/(18+18) = .6111$.

$$\begin{aligned} s_p &= \sqrt{\hat{p}(1-\hat{p})((1/n_1)+(1/n_2))} \\ &= \sqrt{.6111(1-.6111)((1/18)+(1/18))} \\ &= .1625 \end{aligned}$$

$$\begin{aligned} z &= (\hat{p}_1 - \hat{p}_2)/s_p \\ &= (.7222 - .5000)/.1625 \\ &= 1.37 \end{aligned}$$

Using Table A we find $P = 1 - .9147 = .0853$. We do not have sufficient evidence to conclude that the mortality rate is higher on plasterboard.

8.33 (a) $\hat{p}_1 = 72/80 = .90$. $\hat{p}_2 = 59/73 = .8082$. $\hat{p} = (X_1+X_2)/(n_1+n_2) = (72+59)/(80+73) = .8562$.

$$\begin{aligned} s_p &= \sqrt{\hat{p}(1-\hat{p})((1/n_1)+(1/n_2))} \\ &= \sqrt{.8562(1-.8562)((1/80)+(1/73))} \\ &= .0568 \end{aligned}$$

$$z = (\hat{p}_1 - \hat{p}_2)/s_p$$
$$= (.90 - .8082)/.0568$$
$$= 1.62$$

From Table A we find the P-value for a one-sided alternative as $P = 1 - .9474 = .0526$. For our two sided-alternative we double this value, giving $P = .1052$. There is not clear evidence to conclude that the proportions are different. **(b)** In Exercise 8.29 H_0 was clearly rejected. With the smaller sample sizes in this exercise it could not be rejected even though the sample proportions are similar. As the sample size increases the standard error decreases and the z statistic increases. Larger sample sizes give more power.

8.34 (a) $\hat{p} = (X_1 + X_2)/(n_1 + n_2) = (26 + 18)/(36 + 36) = .6111$.

$$s_p = \sqrt{\hat{p}(1-\hat{p})((1/n_1) + (1/n_2))}$$
$$= \sqrt{.6111(1 - .6111)((1/36) + (1/36))}$$
$$= .1149$$

$$z = (\hat{p}_1 - \hat{p}_2)/s_p$$
$$= (.7222 - .5000)/.1149$$
$$= 1.93$$

Using Table A we find $P = 1 - .9732 = .0268$. We conclude that the mortality rate is greater on plasterboard than on glass. **(b)** In Exercise 8.34 we did not reject the null hypothesis. The larger sample sizes produce a smaller standard error. This causes the z to be larger and the P-value to be smaller.

8.35 (a) No, the study was not an experiment because no treatments were imposed. **(b)** The percentage of patients surviving in the group that did not own pets is $28/(11 + 28) = .72$, or 72%; the percentage in the group that owned pets is $50/(50 + 3) = .94$, or 94%. It appears that pet owners have a higher chance of survival. **(c)** The null hypothesis is that the survival rate of patients with pets is the same as the survival rate of patients without pets. The alternative is that the two rates are not equal. **(d)** The value of the test statistic is $X^2 = 8.85$ with $(2-1)(2-1) = 1$ degree of freedom. The P-value is .003. (e) We conclude that pet owners have a higher chance of survival. We cannot conclude that pet ownership is an effective treatment because there was no random assignment of patients to the pet and no pet groups.

8.36 (a) The column percents are 72.92, 27.08, 48.15, 51.85. For the years in which January is up, 72.92% of the time the market is up in the rest of the year, etc. (b) The row percents are 72.92, 27.08, 48.15, 51.85. For the years in which the market is up from February to December, 72.92% of the time the market is up in January, etc. (c) The null hypothesis is that there is no relation between the behavior of the market in January and the behavior in the rest of the year. The alternative is that the January behavior predicts the rest of the year. (d) Reading down columns: 30.72, 17.28, 17.28, 9.72. The expected counts exceed sample counts in the 1,2 and 2,1 cells; they are smaller than the sample counts in the other cells. There are, for example, more years in which January is up and the rest of the year is up, than would be expected under the null hypothesis of no relation. This suggests an association between the two events that is consistent with the January indicator. (e) The test statistic is $X^2 = 4.601$, with 1 degree of freedom. The P-value is .032. Using $\alpha = .05$, we have evidence for rejecting the null hypothesis. If we had used a one-sided test to compare two proportions, we would obtain a P-value one-half as large. (f) The January performance of the market can be used to predict the performance in the rest of the year.

8.37
(a) The data table is

	Reg	WS	All
Hits	2584	35	2619
No Hits	7280	63	7343
Sum	9864	98	9962

(b) The column percents are

	Reg	WS	All
Hits	26.20	35.71	26.29
No Hits	73.80	64.29	73.71

The first row of column percents gives the percentages of hits to at bats. Reggie was a much better hitter in World Series games (37.71% or a 377 "batting average") than in regular season play (26.2% or a 262 "batting average"). **(c)** The null hypothesis is that Jackson's hitting percentage is the same for regular season games and World Series games. The alternative is that the percentages are not equal. To calculate expected counts we use the formula $E_{ij} = R_i C_j / n$. For example, $E_{11} = 2619(9864)/9962 = 2593.2$. The table of expected counts is

	Reg	WS
Hits	2593.2	25.8
No Hits	7270.8	72.2

$X^2 = (2584 - 2593.2)^2/2593.2 + (35 - 25.8)^2/25.8 + (7280 - 7270.8)^2/7270.8 + (63 - 72.2)^2/72.2 = 4.50$. The degrees of freedom are 1. From Table G we find $.025 \leq P \leq .050$. A more accurate computation gives $X^2 = 4.536$ and $P = .033$. Reggie Jackson was a better hitter in World Series play than in the regular season.

8.38 (a) For egg producers 45.94% have mild rodent problems, 44.59% have moderate problems, and 9.46% have severe problems. For turkey producers 45.83% have mild rodent problems, 48.83% have moderate problems, and 8.33% have severe problems. The percents of the different types of problems are very similar for the two types of operation. **(b)** The null hypothesis is that the proportions of the different types of rodent problems are the same for the two types of producers. The alternative is that they are different. $H_0 : p_{ij} = r_i c_j$ for all i and j, $H_a : p_{ij} \neq r_i c_j$ for some i and j. **(c)** The expected counts (reading down columns) are 34.0, 33.4, 6.7; 22.0, 21.6, 4.3. Using formula (8.1), $X^2 = .051$, df=2, $P = .975$. There is no evidence to conclude that the severity of the rodent problem is related to the type of operation.

8.39 (a) The column percentages give the percentages of companies responding and not responding for each size group. They are

	Small	Medium	Large	All
Response	62.5	40.5	20.0	41.0
No Response	37.5	59.5	80.0	59.0

The last column gives the percentages for all companies combined. As the size of the company increases, the response rate decreases. **(b)** The null hypothesis is that the response rate does not depend on the size of the company. The alternative is that there is a relationship between the size of the company and the response rate. **(c)** The table of expected counts is

	Small	Medium	Large
Response	82	82	82
No Response	118	118	118

(e) Using formula (8.1), $X^2 = 74.7$. $df = 2$. From Table G we conclude that $P < .0005$. The response rate decreases as the size of the company increases.

8.40 (a) The row sums are 327 and 443; The column sums are 214, 226, and 330. **(b)**

	Urban	Intermediate	Rural	All
Tetracycline	30.37	39.82	52.12	42.47
No Tetracycline	69.63	60.18	47.88	57.53

The percentage of physicians prescribing tetracycline increases as the county type varies from urban to intermediate to rural. **(c)** The null hypothesis states that there is no difference in the proportion of physicians prescribing tetracycline in the three county types. The alternative is that the three county types are not all the same in this regard. $H_0 : p_{ij} = r_i c_j$ for all i and j, $H_a : p_{ij} \neq r_i c_j$ for some i and j. **(d)** The expected counts (reading down columns) are 90.9, 123.1; 96.0, 130.0; 140.1, 189.9. The test statistic is $X^2 = 26.04$ with 2 degrees of freedom. From Table G, we find $P < .001$. There is evidence to conclude that the location of the practice is related to whether tetracycline is prescribed. Physicians in rural counties are most likely to prescribe it. Those in intermediate counties are less likely and those in urban counties are least likely.

8.41 (a) The table of counts with row and column sums is

	Family	Pediatrics	Other	All
Pet	327	32	159	518
No Pet	443	122	808	1373
Sum	770	154	967	1891

(b) The column percents give the percents of physicians who prescribe tetracycline and who do not for each type of practice. They are given in the following table with the percents for all physicians combined in the last column.

	Family	Pediatrics	Other	All
Pet	42.47	20.78	16.44	27.39
No Pet	57.53	79.22	83.56	72.61

The family practitioners are the most likely prescribe. The pediatricians are less likely and the other types of physicians are least likely to prescribe tetracycline. **(c)** The null hypothesis is that there are no differences between physicians in different types of practice in regard to their likelihood of prescribing tetracycline. The alternative is that the three groups of physicians are not all the same in this regard. $H_0 : p_{ij} = r_i c_j$ for all i and j, $H_a : p_{ij} \neq r_i c_j$ for some i and j. The table of expected counts is

	Family	Pediatrics	Other
Pet	210.9	42.2	264.9
No Pet	559.1	111.8	702.1

Using formula (8.1), we calculate $X^2 = 149.66$ with 2 degrees of freedom. From Table G we conclude $P < .0005$. The proportion of physicians prescribing tetracycline varies with the type of practice. The family practitioners have a very high rate compared to the others. The other category of physicians has the lowest rate with pediatricians in the middle.

8.42 The Column percents are 34.54, 19.08, 27.63, 18.75; 10.77, 7.69, 56.92, 24.62; 13.25, 15.66, 50.60, 20.48. For women with higher nicotine consumption there is a tendency for greater alcohol consumption. The row percents are 85.37, 5.69, 8.94; 76.32, 6.58, 17.11; 51.53, 22.70, 25.77; 63.33, 17.78, 18.89. For women with a higher alcohol consumption there is a higher nicotine consumption. The null hypothesis is that there is no relation between alcohol consumption and nicotine consumption. The alternative is that these two variables are related. $H_0 : p_{ij} = r_i c_j$, $H_a : p_{ij} \neq r_i c_j$ for some i and j. The expected counts (reading down columns) are 82.7, 51.1, 109.6, 60.5; 17.7, 10.9, 23.4, 12.9; 22.6, 14.0, 29.9, 16.5. The test statistic is $X^2 = 42.25$ with 6 degrees of freedom. From Table G we find $P < .001$. There is evidence to conclude that there is a relationship between alcohol consumption and nicotine consumption. Higher alcohol consumption is associated with higher nicotine consumption.

8.43 The data table is

	Normal	I	II	III-IV	All
URI	95	143	144	70	452
Diarrhea	53	94	101	48	296
URI and diarrhea	27	60	76	27	190
None	113	48	44	22	227
Sum	288	345	365	167	1165

The table of column percents is

	Normal	I	II	III-IV
URI	32.99	41.45	39.45	41.92
Diarrhea	18.40	27.25	27.67	28.74
URI and diarrhea	9.38	17.39	20.82	16.17
None	39.24	13.91	12.05	13.17

The three nutritionally inadequate groups have similar profiles. The normal group has lower rates for each of the three categories of disease. The table of row percents is

	Normal	I	II	III-IV
URI	21.02	31.64	31.86	15.49
Diarrhea	17.91	31.76	34.12	16.22
URI and diarrhea	14.21	31.58	40.00	14.21
None	49.78	21.15	19.38	9.69

The percents in the different nutritional status categories appear to vary with the illness group. The most noticeable difference is between the no illness category and the three others. Approximately half of the children with no illness were in the normal nutritional group. The proportion of children with normal nutritional status and no illness is $113/1165 = 9.7$ percent. This percent would be found in a table of total or overall percents. The null hypothesis is that there is no association between nutritional status and the likelihood of illness. The alternative is that these two variables are associated. $H_0 : p_{ij} = r_i c_j$ for all i and j, $H_a : p_{ij} \neq r_i c_j$ for some i and j. The table of expected counts is

	Normal	I	II	III-IV
URI	111.7	133.9	141.6	64.8
Diarrhea	73.2	87.7	92.7	42.4
URI and diarrhea	47.0	56.3	59.5	27.2
None	56.1	67.2	71.1	32.5

Using formula (8.1), we calculate the test statistic $X^2 = 101.29$, with 9 degrees of freedom. From Table G we find $P < .0005$. We conclude that there is a relationship between nutritional status and illness. In general, we see more illness associated with poorer nutritional status.

8.44 The conclusion is not justified because this is a case of voluntary response. The callers are not a random sample of citizens in general.

8.45 The modification is effective in reducing the defective rate from 11% to 5%. We want to test $H_0 : p = .11$ versus $H_a : p < .11$. The sample proportion of defectives is $\hat{p} = 16/300 = .0533$. The test statistic is $z = (\hat{p} - p_0)/\sqrt{p_0(1-p_0)/n} = (.0533 - .11)/\sqrt{.11(1-.11)/300} = -3.13$. Using Table A we find that $P = .0009$. We assume that the old rate is known to be 11% and that the number is defectives in the new sample is binomial.

8.46 The 95% confidence interval is $\hat{p} \pm z^*\sqrt{\hat{p}(1-\hat{p})/n} = .0533 \pm 1.96\sqrt{.0533(1-.0533)/300} = (.0278, .0788)$. To construct a 95% confidence interval for $p - p_0$ we simply subtract .11 from the lower and upper limits of the confidence interval for p. The interval is $(-.0822, -.0312)$. There is a reduction of between 3.12% and 8.22% in the percentage of defectives.

8.47 (a) The sample proportion is $\hat{p} = 444/950 = .4674$. The 99% confidence interval is $\hat{p} \pm z^*\sqrt{\hat{p}(1-\hat{p})/n} = .4674 \pm 2.576\sqrt{.4674(1-.4674)/950} = (.428, .512)$. (b) In terms of percent the interval is 42.8% to 51.2%. (c) To convert the interval to a number of students, we multiply the lower and upper limits by the number of undergraduates (35,000). The interval is (14980, 17920). This answer depends on how the original interval was rounded.

8.48 (a) The proportion of women in the freshman class is $p_0 = 214/851 = .2515$. (b) The proportion of women in the top 30 is $\hat{p} = 15/30 = .5$. (c) The hypotheses are $H_0 : p = .2515$ and $H_a : p \neq .2515$. The test statistic is $z = (\hat{p} - p_0)/\sqrt{p_0(1-p_0)/n} = (.5 - .2515)/\sqrt{.2515(1-.2515)/30} = 16.71$. From Table A we find $P = 2(1 - .9998) = .0004$. A higher proportion of the women are among the top 30 students than would be expected.

8.49 We perform a significance test for a single proportion. The hypotheses are $H_0 : p = .5$, $H_a : p \neq .5$. The sample proportion is $\hat{p} = .75$. The test statistic is $z = (\hat{p} - p_0)/\sqrt{p_0(1-p_0)/n} = (.75 - .5)/\sqrt{.5(1-.5)/20} = 2.24$. From Table A we find $P = 2(1 - .9875) = .0250$. H_0 is rejected. There is evidence to conclude that the proportion in Tippecanoe county differs from the national average. The rate appears to be higher. Before looking at the data we would have no reason to hypothesize a particular one-sided alternative.

8.50 47% of the blacks are vegetarians, and 61% of the whites are vegetarians. We test the null hypothesis that the proportions are the same versus the two-sided alternative. $\hat{p} = (X_1 + X_2)/(n_1 + n_2) = (42 + 135)/(89 + 223) = .5673$. $\hat{p}_1 = 42/89 = .4719$. $\hat{p}_2 = 135/223 = .6054$.

$$\begin{aligned} s_p &= \sqrt{\hat{p}(1-\hat{p})((1/n_1) + (1/n_2))} \\ &= \sqrt{.5673(1 - .5673)((1/89) + (1/223))} \\ &= .06212 \end{aligned}$$

$H_0 : p_1 = p_2$, $H_a : p_1 \neq p_2$.

$$z = (\hat{p}_1 - \hat{p}_2)/s_p$$
$$= (.4719 - .6054)/.06212$$
$$= -2.15$$

From Table A we find $P = 2(.0158) = .0316$. A more accurate calculation gives $P = .0317$. There is evidence to conclude that the proportion of vegetarians among blacks is higher for the people attending this meeting. Inferences to Seventh Day Adventists in general and blacks and whites in general are beyond the realm of this particular study.

8.51 (a) High or low blood pressure is the explanatory variable. The sample proportions of deaths for the two groups are $\hat{p}_1 = 21/2676 = .0078$ and $\hat{p}_2 = 55/3338 = .0165$, where group one is the low blood pressure group and group two is the high blood pressure group. **(b)** The hypotheses are $H_0 : p_1 = p_2$ and $H_a : p_1 < p_2$. For the combined data, $\hat{p} = (X_1 + X_2)/(n_1 + n_2) = (21 + 55)/(2676 + 3338) = .0126$.
The standard error is

$$s_p = \sqrt{\hat{p}(1-\hat{p})((1/n_1) + (1/n_2))}$$
$$= \sqrt{.0126(1-.0126)((1/2676) + (1/3338))}$$
$$= .0029$$

and the test statistic is

$$z = (\hat{p}_2 - \hat{p}_1)/s_p$$
$$= (.0165 - .0078)/.0029$$
$$= 2.98$$

From Table A we find $P = 1 - .9986 = .0014$. We conclude that high blood pressure is associated with an increased risk. **(c)**

	Low	High
Dead	21	55
Not dead	2655	3283

CHAPTER 8

The chi-square test is not appropriate for this problem because we have a one-sided alternative. It is appropriate only for a two-sided alternative. **(d)**

$$s_D = \sqrt{(\hat{p}_1(1-\hat{p}_1)/n_1) + (\hat{p}_2(1-\hat{p}_2)/n_2)}$$
$$= \sqrt{(.0078(1-.0078)/2676) + (.0165(1-.0165)/3338)}$$
$$= .0028$$

$$\hat{p}_2 - \hat{p}_1 \pm z^* s_D = .0165 - .0078 \pm 1.960(.0028)$$
$$= (.0032, .0141)$$

8.52 (a) The hypotheses are $H_0 : p_1 = p_2$ and $H_a : p_1 < p_2$. For the combined data, $\hat{p} = (X_1 + X_2)/(n_1 + n_2) = (457 + 437)/(1003 + 620) = .5508$. The two sample proportions are $\hat{p}_1 = 457/1003 = .4556$ and $\hat{p}_2 = 437/620 = .7048$. The standard error is

$$s_p = \sqrt{\hat{p}(1-\hat{p})((1/n_1) + (1/n_2))}$$
$$= \sqrt{.5508(1-.5508)((1/1003) + (1/620))}$$
$$= .0254$$

The test statistic is

$$z = (\hat{p}_1 - \hat{p}_2)/s_p$$
$$= (.4556 - .7048)/.0254$$
$$= -9.81$$

From Table A we find $P < .0002$. The program has increased the proportion of mothers who follow the practice of feeding children with diarrhea. **(b)** The chi-square test in inappropriate here because we have a one-sided alternative. The chi-square statistic can be used when we have a two-sided alternative. **(c)**

$$s_D = \sqrt{(\hat{p}_1(1-\hat{p}_1)/n_1) + (\hat{p}_2(1-\hat{p}_2)/n_2)}$$
$$= \sqrt{(.4556(1-.4556)/1003) + (.7048(1-.7048)/620)}$$
$$= .0241$$

$$\hat{p}_1 - \hat{p}_2 \pm z^* s_D = .4556 - .7048 \pm 1.960(.0241)$$
$$= (-.297, -.202)$$

8.53 There is no evidence to conclude that there is an association between the use of aluminum–containing antacids and Alzheimer's disease. The test statistic is $X^2 = 7.118$, with 3 degrees of freedom. The P-value is .068.

8.54 We will use column percents to describe the data. These give the percents of students in the three status categories separately for men and women.

Status	Men	Women
Completed	53.21	42.79
Still enrolled	16.86	14.41
Dropped out	29.94	42.79

The null hypothesis is that the proportions of men in the three status categories are the same as the proportions of women. The alternative is that they are not the same. The test statistic is $X^2 = 13.398$ with 2 degrees of freedom. The P-value is reported by software as .001. We conclude that that men and women are distributed differently in the three status categories. Men are more likely to complete their degrees and women are more likely to drop out. Other factors that might be relevant to this study include field of study, financial support and part–time versus full–time status.

8.55 The data table is

	Ireland	Portugal	Norway	Italy	All
Tasters	558	345	185	402	1490
Non-tasters	225	109	81	134	549
Sum	783	454	266	536	2039

The table of row percents is

	Ireland	Portugal	Norway	Italy
Tasters	37.45	23.15	12.42	26.98
Non-tasters	40.98	19.85	14.75	24.41
Sum	38.40	22.27	13.05	26.29

The percent of tasters appears to vary with country. We test the null hypothesis $H_0 : p_{1(1)} = p_{1(2)} = p_{1(3)} = p_{1(4)}$ and $p_{2(1)} = p_{2(2)} = p_{2(3)} = p_{2(4)}$ versus the alternative H_a : at least one of the equalities in H_0 does not hold. Formula (8.2) gives the test statistic as $X^2 = 5.96$. Using Table G with three degrees of freedom we find $.10 \leq P \leq .15$. The exact value is $P = .114$. The variation between countries seen in the table of row percents is not distinguishable from chance. We have no evidence to conclude that the proportions of PTC tasters is different in these four countries.

8.56 The column percentages are: Hawaiians 40.75, 53.32, 3.81, 2.12. Hawaiian-white 44.77, 46.79, 6.07, 2.36. Hawaiian-Chinese 40.97, 43.97, 10.55, 4.51. White 43.00, 40.00, 13.00, 4.00. We test the null hypothesis $H_0 : p_{1(1)} = p_{1(2)} = p_{1(3)} = p_{1(4)}$ and $p_{2(1)} = p_{2(2)} = p_{2(3)} = p_{2(4)}$ and $p_{3(1)} = p_{3(2)} = p_{3(3)} = p_{3(4)}$ and $p_{4(1)} = p_{4(2)} = p_{4(3)} = p_{4(4)}$ versus the alternative H_a : at least one of the equalities in H_0 does not hold. Formula (8.2) gives the test statistic as $X^2 = 1078.60$. Using Table G with nine degrees of freedom we find $P < .0005$. There is evidence to conclude that the proportions of the four blood types are not the same in all four ethnic groups.

8.57 The data table for the British study is

	Aspirin	No Aspirin	All
Death	148	79	227
No Death	3281	1631	4912
Sum	3429	1710	5139

The data table for the American study is

	Aspirin	No Aspirin	All
Death	104	189	293
No Death	10933	10845	21778
Sum	11037	11034	22071

The table of column percents for the British study is

	Aspirin	No Aspirin	All
Death	4.32	4.62	4.42
No Death	95.68	95.38	95.58

The table of column percents for the American study is

	Aspirin	No Aspirin	All
Death	.94	1.71	1.33
No Death	99.06	98.29	98.67

We test $H_0: p_{1(1)} = p_{1(2)}$ and $p_{2(1)} = p_{2(2)}$ versus the alternative H_a: at least one of the equalities in H_0 does not hold. Using formula (8.2) we calculate $X^2 = .25$ with one degree of freedom for the British study. From Table G we find $P > .25$. The exact value is $P = .618$. For the American study, $X^2 = 25.01$ with one degree of freedom. From Table G we find $P < .0005$. The positive effects of aspirin on cardiovascular disease were demonstrated in the American study but not in the British study. The American study used a much larger sample size, the duration was one year shorter and the aspirin was taken every other day rather than every day. These analyses could be performed as a comparison of two proportions. The conclusions would be the same.

8.58 (a) $H_0: p_1 = p_2$, $H_a: p_1 \neq p_2$. $\hat{p} = (X_1 + X_2)/(n_1 + n_2) = (28 + 30)/(82 + 78) = .3625$.
$\hat{p}_1 = 28/82 = .3415$. $\hat{p}_2 = 30/78 = .3846$.

$$\begin{aligned} s_p &= \sqrt{\hat{p}(1-\hat{p})((1/n_1) + (1/n_2))} \\ &= \sqrt{.3625(1-.3625)((1/82) + (1/78))} \\ &= .0760 \end{aligned}$$

$$\begin{aligned} z &= (\hat{p}_1 - \hat{p}_2)/s_p \\ &= (.3415 - .3846)/.0760 \\ &= -.5671 \end{aligned}$$

From Table A we find $P = 2(.2843) = .57$. (b)

	Gastric Freezing	Placebo	All
Improved	28	30	58
No Improvement	54	48	102
Sum	82	78	160

$H_0: p_{1(1)} = p_{1(2)}$ and $p_{2(1)} = p_{2(2)}$. The alternative is that one of the equalities in the null hypothesis is false. $X^2 = .322$, $P = .57$. $z^2 = (-.5671)^2 = .321$. The difference is due to roundoff. (c) Gastric freezing is not an effective treatment for ulcers.

8.59 (a) The data table is

	Hospital A	Hospital B	All
Death	63	16	79
No Death	2037	784	2821
Sum	2100	800	2900

The test statistic is $X^2 = 2.19$ with one degree of freedom. From Table G we find $.10 \leq P \leq .15$. The exact value is $P = .139$. There is no evidence to conclude that the death rates in the two hospitals are different. (b) The data table for patients in good condition is

	Hospital A	Hospital B	All
Death	6	8	14
No Death	594	592	1186
Sum	600	600	1200

The test statistic is $X^2 = .29$ with one degree of freedom. From Table G we find $P > .25$. The exact value is $P = .591$. There is no evidence to conclude that the death rates for patients in good condition in the two hospitals are different. The data table for patients in poor condition is

	Hospital A	Hospital B	All
Death	57	8	65
No Death	1443	192	1635
Sum	1500	200	1700

The test statistic is $X^2 = .02$ with one degree of freedom. From Table G we find $P > .25$. The exact value is $P = .890$. There is no evidence to conclude that the death rates in the two hospitals are different. (c) Although the sample proportions illustrate Simpson's paradox as described in Example 2.25, none of the differences are statistically significant.

8.60 $\hat{p} = .5$. (n, z, P)=(15, 1.10, .27), (25, 1.41, .16), (50, 2.00, .05), (75, 2.45, .014), (100, 2.83, .005), (500, 6.32, .0000).

8.61 For all of the intervals $z^* = 1.96$ and

$$\begin{aligned} s_D &= \sqrt{(\hat{p}_1(1-\hat{p}_1)/n_1) + (\hat{p}_2(1-\hat{p}_2)/n_2)} \\ &= \sqrt{(.6(1-.6)/n) + (.4(1-.4)/n)} \end{aligned}$$

Therefore, the margins of error are $1.96 s_D$.

n	15	25	50	75	100	500
s_D	.18	.14	.10	.08	.07	.03
m	.35	.27	.19	.16	.14	.06

8.62 $(n$, margin of error$)$= (10, .58), (30, .33), (50, .26), (100, .18), (200, .13), (500, .08).

8.63 (a) and (b)

$$\begin{aligned} s_D &= \sqrt{(\hat{p}_1(1-\hat{p}_1)/n_1) + (\hat{p}_2(1-\hat{p}_2)/n_2)} \\ &= \sqrt{.5^2/n + .5^2/n} \end{aligned}$$

Since $m = z^* s_D$, it follows that $n = (z^*)^2/2m^2$. For a 95% confidence interval with $m = .05$ we calculate $n = 1.96^2/2(.05)^2 = 768.32$. A sample of size 769 is needed.

8.64 It is not possible to find the value of n_2 to guarantee the desired result. Even if n_2 is so large that the term $.25/n_2$ can be neglected, then $s_D = \sqrt{.0125}$ and $m = .1839$ which is greater than .1.

8.65 Here is a segmented bar chart of the data.

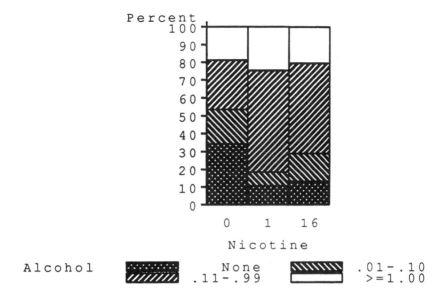

8.66 Here is a segmented bar chart of the data.

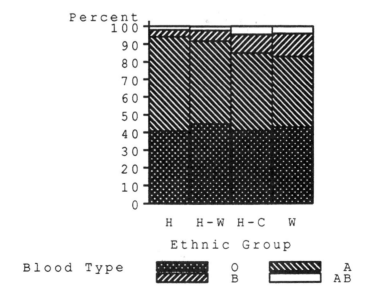

8.67 The death rates are so low that these small sample sizes are very unlikely to show any difference between the groups.

8.68 The data table for social comparison is

	Male	Female	All
High	49	21	70
Low	18	46	64
Sum	67	67	134

The table of column percents for social comparison is

	Male	Female	All
High	73.13	31.34	52.24
Low	26.87	68.66	47.76

It appears that a much larger percentage of males than females have high social comparison as a goal. This is supported by the analysis. The test statistic is $X^2 = 23.45$ with one degree of freedom. From Table G we find $P < .0005$.

The data table for mastery is

	Male	Female	All
High	36	35	71
Low	31	32	63
Sum	67	67	134

The table of column percents for mastery is

	Male	Female	All
High	53.73	52.24	52.99
Low	46.27	47.76	47.01

The percentages of students in the high mastery group are very similar for males and females. This is supported by the analysis. The test statistic is $X^2 = .03$ with one degree of freedom. From Table G we find $P > .25$. The exact value is $P = .863$. Therefore, it appears that the effects found for these data in the text were due to differences in social comparison rather than mastery.

8.69 The percentages of students having loans in the different fields are 47.76, 42.53, 41.70, 41.78, 32.00, 51.67, 44.53. The differences are not statistically significant ($X^2 = 6.525$ with 6 degrees of freedom; $P = .367$).

8.70 There appears to be a relation between the PEOPLE score and field of study. The test results are $X^2 = 43.487$ with 12 degrees of freedom and $P < .0005$. Fields with a relatively large proportion of high scores are Child Development and Family Studies and Liberal Arts. Fields with a relatively large proportion of high scores are Management, Engineering and Science.

8.71 The death percentages are .44, .52, .14, .16, .29, .27, .60, .85. A plot of these percentages shows the relationship between age and death rate. The test statistic is $X^2 = 19.715$ with 7 degrees of freedom. The P-value is .006. We conclude that survival and age are related. The death rates are relatively high for children under 4 and for those over 30. The data do not provide any information regarding the chances of catching measles.

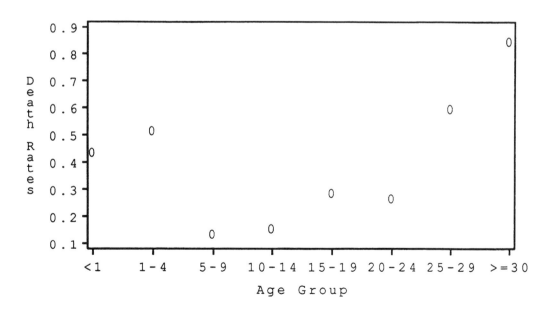

8.72 The data table for hypertension is

	Hypo	Normal	Hyper
Yes	966	3662	11
No	128	1027	16

There is an association between hypertension and potassium status ($X^2 = 83.147$ with 2 degrees of freedom and $P < .0005$). The reanalysis without the hyperkalemic group gives $X^2 = 57.764$ with 1 degrees of freedom and $P < .0005$.

The data table for heart failure is

	Hypo	Normal	Hyper
Yes	181	1158	15
No	913	3531	12

There is an association between heart failure and potassium status ($X^2 = 48.761$ with 2 degrees of freedom and $P < .0005$). The reanalysis without the hyperkalemic group gives $X^2 = 33.125$ with 1 degrees of freedom and $P < .0005$.

The data table for diabetes is

	Hypo	Normal	Hyper
Yes	225	1196	8
No	869	3493	19

There is an association between diabetes and potassium status ($X^2 = 12.042$ with 2 degrees of freedom and $P = .002$). The reanalysis without the hyperkalemic group gives $X^2 = 11.678$ with 1 degrees of freedom and $P = .001$.

The data table for sex is

	Hypo	Normal	Hyper
Female	793	3189	13
Male	301	1500	14

There is an association between sex and potassium status; females are more likely to have low potassium levels ($X^2 = 13.639$ with 2 degrees of freedom and $P = .001$). The reanalysis without the hyperkalemic group gives $X^2 = 8.288$ with 1 degrees of freedom and $P = .004$.

All four of the risk factors are associated with potassium status. A more advanced analysis of these data would simultaneously examine the effects of these variables.

8.73 The percentages of women by year are 23.43, 27.86, 32.29. 40.43, 43.14, 49.00, 52.71, 54.86, 58.86. The proportion of women enrolled in pharmacy programs has increased over time. The test statistic is $X^2 = 359.186$, with 8 degrees of freedom. The P-value is less than .0005. The plot of the percentages is roughly linear. The least squares line is $\hat{y} = -4443 + 2.27x$ We would not be willing to use the line to predict the percentage of women pharmacy students in the year 2000. This would require extrapolation far beyond the range of the data available. The predicted value from the least squares line is 97%.

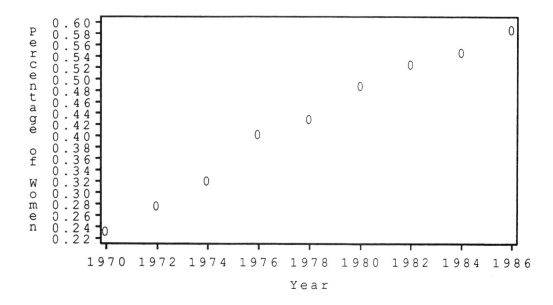

8.74 (a) The proportion of eligible jurors that are Mexican Americans is $p_0 = 143611/181535 = .7911$. **(b)** The proportion of the selected jurors who are Mexican American is $\hat{p} = 339/870 = .3897$. The test indicates that Mexican Americans are underrepresented on these juries ($z = 29.1965$, $P < .0005$). Note we could do a one-sided or two-sided test here; the conclusion is the same. **(c)** For the two sample problem we compare the proportion of Mexican Americans among those selected with the proportion among those not selected. The two sample proportions are $\hat{p}_1 = 339/870 = .3897$ and $\hat{p}_2 = 143272/180665 = .7930$. The test statistic is $z = 29.20$ with $P < .0005$. The conclusion is the same. (c) Here is the data table

	MA	other
Selected	339	531
Not selected	143272	37393

The test statistic is $X^2 = 852.433$ with 1 degree of freedom and $P < .0005$. The square of the z statistic equals X^2. Any differences are due to round off error.

4.9 CHAPTER 9

9.1 (a) The plot indicates a fairly strong straight line pattern. The value of r^2 is .886 or 88.6%. The observation with a moderately large residual is 1983.

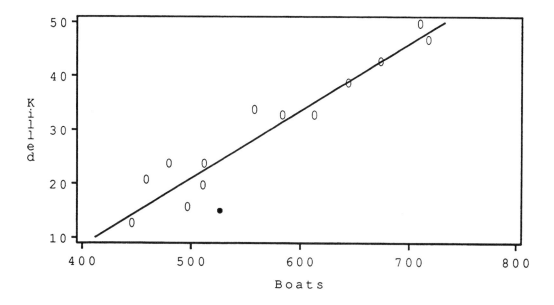

(b) The regression equation is $\hat{y} = -41.4 + 0.125x$. The hypotheses are $H_0 : \beta_1 = 0$ and $H_a : \beta_1 > 0$. The t statistic is 9.68 with 12 degrees of freedom. The P-value is less than .0005. There is strong evidence to conclude that there is a positive linear association between powerboat registrations and manatees killed.

9.2 (a) The circled points do not deviate from the linear pattern of the other points. It appears that the state's actions have not reduced the number of manatees killed.

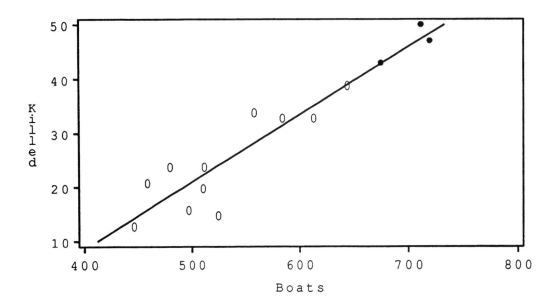

(b) The predicted number of manatees killed corresponding to 700,000 registered powerboats is 45.97. The prediction interval is (35.63, 56.31). These values can be read from the output. A prediction interval, rather than a confidence interval, is preferred because we are trying to predict the number of manatees killed in a future year (with 700,000 powerboat registrations) rather than the mean number of kills for many years with this number of registrations.

9.3 (a) There appears to be a strong linear pattern.

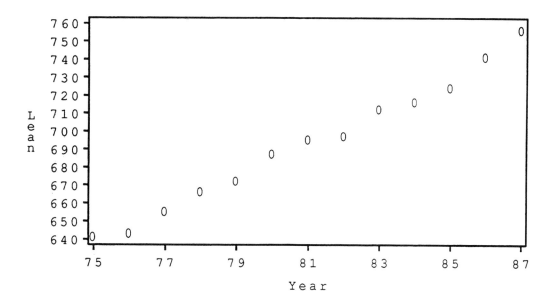

(b) The equation of the least-square line is $\hat{y} = -61.12 + 9.31x$. The line explains 98.8% of the variation in lean. (c) A 95% confidence interval for the average rate of change of the lean (the slope of the true regression line) is $b_1 \pm t^* s_{b_1} = 9.32 \pm 2.201(.3099)$ or $(8.64, 10.00)$. Here we used $t^* = 2.201$, the value corresponding to 11 degrees of freedom in Table E.

9.4 (a) The predicted value is $\hat{y} = -61.12 + 9.31(18) = 107$. This corresponds to a lean of 2.9107 (for our coding we take 2.9 and add the last three digits from the prediction equation). (b) The residual for 1918 is $2.9071 - 2.9107 = -.0036$. The coded value is -36. This residual is much larger than any of the residuals for 1975 to 1987. This fact can be easily seen in the following plot of all the data with the regression line computed from 75 to 87.

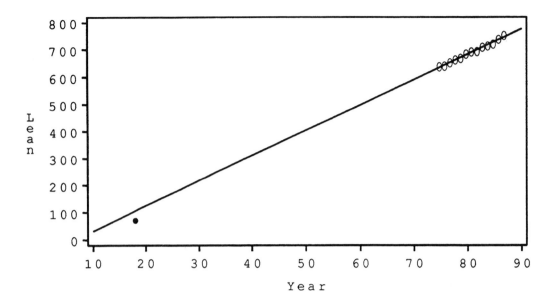

9.5 (a) The predicted value from the equation is $\hat{y} = -61.12 + 9.31(97) = 843$. The lean is 2.9 plus .0843 or 2.9843. **(b)** A prediction interval is preferred because you want to predict the value of the lean in the future year 1997.

9.6 (a) The percentage of variation in attendance explained by ticket price is .01%. **(b)** The hypotheses are $H_0 : \beta_1 = 0$ and $H_a : \beta_1 \neq 0$. The t statistic is .046 with 25 degrees of freedom. The P-value is .9637. The linear regression is of no value in this situation. **(c)** We should always plot the data. The plot may lead to other approaches to the analysis of the data.

9.7 (a)

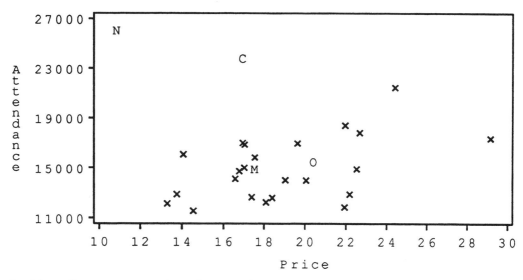

C=Charlotte M=Miami N=Minnesota O=Orlando X=Others

(b) Minnesota and Charlotte have high values for attendance. Both are outliers and influential. Minnesota is much more influential than Charlotte. Miami and Orlando fit within the general linear pattern of the rest of the data. (c) The regression equation is $\hat{y} = 8962 + 317x$. The value of r^2 is .23. $F = 6.945$ and $P = .0148$. The least squares line has changed substantially. The value of r^2 is much larger and the linear relationship is now statistically significant. (d) The results of these analyses show an association with teams as the unit of analysis. Teams with higher priced tickets tend to have greater attendance. The two outliers have very high attendance. The results do not indicate how attendance might change if ticket prices would change. Lurking variables would include the team performance and cost of living as reflected in prices for entertainment.

9.8 (a) The slope of the population regression line represents the increase natural gas consumption corresponding to an increase of one heating degree day. The degrees of freedom are $n - 2 = 16$. From Table E we find $t^* = 2.120$ for a 95% confidence interval. $b_1 \pm t^* s_{b_1} = .26896 \pm 2.120(.00815) = (.252, .286)$. (b) The margin of error of this interval should be smaller. The margin of error in Example 9.9 is .02710. and the margin of error for the interval we calculated here is $2.120(.00815) = .01730$. This confirms our expectation.

9.9 (a) The intercept represents gas consumption for uses other than heating. The degrees of freedom are $n - 2 = 16$. From Table E we find $t^* = 2.120$ for a 95% confidence interval. $b_0 \pm t^* s_{b_0} = 2.405 \pm 2.120(.20351) = (1.9736, 2.8364)$. **(b)** The width is .8628. It is shorter than the interval in Example 9.5 because it is based on a larger sample. The margins of error are half of the widths and are related in the same way.

9.10 (a) $t = b_1/s_{b_1} = .26896/.00815 = 33.00$. **(b)** The degrees of freedom are 16. From Table E we find the critical value for $\alpha = .05$ test to be 1.746. Since our calculated value of t is greater than this critical value, we reject the null hypothesis at the 5% level. **(c)** $P < .0005$.

9.11 We test the following hypotheses $H_0 : \beta_1 = 0$. $H_a : \beta_1 \neq 0$. The test statistic is $t = b_1/s_{b_1} = .82/.38 = 2.16$. The degrees of freedom are $n - 2 = 53$. Using Table E with $df = 60$ we find $.02 \leq P \leq .04$ for the one-sided test. The exact value is $P = .035$. A one-sided test is slightly preferred. We conclude that there is a positive relationship between the pretest and the final exam score.

9.12 (a) The estimate of the mean is $\hat{\mu}_y = b_0 + b_1(30) = 1.233 + .20221(30) = 7.299$. The standard error is

$$s_{\hat{\mu}} = s\sqrt{(1/n) + ((x^* - \bar{x})^2 / \sum(x_i - \bar{x})^2)}$$
$$= .435\sqrt{(1/9) + ((30 - 21.54)^2/1440.646)}$$
$$= .174$$

The degrees of freedom are $n - 2 = 7$. From Table E we find $t^* = 3.499$ for a 99% confidence interval. The interval is $\hat{\mu}_y \pm t^* s_{\hat{\mu}} = 7.299 \pm 3.499(.174) = (6.69, 7.91)$. **(b)** The estimate of the mean is $\hat{\mu}_y = b_0 + b_1(80) = 1.233 + .20221(80) = 17.410$. The standard error is

$$s_{\hat{\mu}} = s\sqrt{(1/n) + ((x^* - \bar{x})^2 / \sum(x_i - \bar{x})^2)}$$
$$= .435\sqrt{(1/9) + ((80 - 21.54)^2/1440.646)}$$
$$= .686$$

The degrees of freedom are $n - 2 = 7$. From Table E we find $t^* = 3.499$ for a 99% confidence interval. The interval is $\hat{\mu}_y \pm t^* s_{\hat{\mu}} = 17.410 \pm 3.499(.686) =$

(15.01, 19.81). **(c)** The margin of error for 80 degree days is larger because 80 is farther from $\bar{x}$ than 30. **(d)** Yes. 80 degree days is much higher than any value in the data set.

9.13 The estimate of the mean is $\hat{\mu}_y = b_0 + b_1(31.54) = 1.233 + .20221(31.54) = 7.611$. The standard error for the confidence interval is

$$s_{\hat{\mu}} = s\sqrt{(1/n) + ((x^* - \bar{x})^2 / \sum(x_i - \bar{x})^2)}$$
$$= .435\sqrt{(1/9) + ((31.54 - 21.54)^2/1440.646)}$$
$$= .185$$

The degrees of freedom are $n - 2 = 7$. From Table E we find $t^* = 1.895$ for a 90% confidence interval. The confidence interval is $\hat{\mu}_y \pm t^* s_{\hat{\mu}} = 7.611 \pm 1.895(.185) = (7.26, 7.96)$.
The standard error for the prediction interval is

$$s_{\hat{\mu}} = s\sqrt{(1 + 1/n) + ((x^* - \bar{x})^2 / \sum(x_i - \bar{x})^2)}$$
$$= .435\sqrt{(1 + 1/9) + ((31.54 - 21.54)^2/1440.646)}$$
$$= .473$$

The degrees of freedom and t^* are the same as above. The prediction interval is $\hat{\mu}_y \pm t^* s_{\hat{\mu}} = 7.611 \pm 1.895(.473) = (6.72, 8.51)$. The margin of error for the prediction interval is larger (.896 versus .351). The 95% intervals would be longer.

9.14 (a) The estimate of the response is $\hat{y} = b_0 + b_1(30) = 1.233 + .20221(30) = 7.299$. The standard error is

$$s_{\hat{y}} = s\sqrt{1 + (1/n) + ((x^* - \bar{x})^2 / \sum(x_i - \bar{x})^2)}$$
$$= .435\sqrt{1 + (1/9) + ((30 - 21.54)^2/1440.646)}$$
$$= .469$$

The degrees of freedom are $n - 2 = 7$. From Table E we find $t^* = 3.499$ for a 99% prediction interval. The interval is $\hat{y} \pm t^* s_{\hat{y}} = 7.299 \pm 3.499(.469) =$

(5.66, 8.94). **(b)** The estimate of the response is $\hat{y} = b_0 + b_1(80) = 1.233 + .20221(80) = 17.410$. The standard error is

$$s_{\hat{y}} = s\sqrt{1 + (1/n) + ((x^* - \bar{x})^2/\sum(x_i - \bar{x})^2)}$$
$$= .435\sqrt{1 + (1/9) + ((80 - 21.54)^2/1440.646)}$$
$$= .812$$

The degrees of freedom are $n - 2 = 7$. From Table E we find $t^* = 3.499$ for a 99% prediction interval. The interval is $\hat{y} \pm t^* s_{\hat{y}} = 17.410 \pm 3.499(.812) = (14.57, 20.25)$. **(c)** The margins of error for the prediction intervals are larger than for the confidence intervals.

9.15 (a) When the price of textiles increases by one unit, the consumption increases by β_1. Since this quantity is negative, we can say the consumption decreases by $-\beta_1$. The degrees of freedom are $n - 2 = 15$. From Table E we find $t^* = 1.753$ for a 90% confidence interval. $b_1 \pm t^* s_{b_1} = -1.3233 \pm 1.753(.1163) = (-1.53, -1.12)$. **(b)** The intercept is of no interest because a price of zero is not meaningful. **(c)** The estimate of the mean is $\hat{\mu}_y = b_0 + b_1(100) = 235.4897 - 1.323306(100) = 103.16$. Using the hint, we calculate $\sum(x - \bar{x})^2 = (n-1)s_x^2 = 16(284.47) = 4551.52$. The standard error is

$$s_{\hat{\mu}} = s\sqrt{(1/n) + ((x^* - \bar{x})^2/\sum(x_i - \bar{x})^2)}$$
$$= 7.84818\sqrt{(1/17) + ((100 - 76.3118)^2/4551.52)}$$
$$= 3.3491$$

The degrees of freedom are $n - 2 = 15$. From Table E we find $t^* = 1.753$ for a 90% confidence interval. The interval is $\hat{\mu}_y \pm t^* s_{\hat{\mu}} = 103.16 \pm 1.753(3.3491) = (97.29, 109.03)$. **(d)** The estimate of the response is $\hat{y} = b_0 + b_1(75) = 235.4897 - 1.3233(75) = 136.24$. The standard error is

$$s_{\hat{y}} = s\sqrt{1 + (1/n) + ((x^* - \bar{x})^2/\sum(x_i - \bar{x})^2)}$$
$$= 7.84818\sqrt{1 + (1/17) + ((75 - 76.3118)^2/4551.52)}$$
$$= 8.0772$$

The degrees of freedom are $n - 2 = 15$. From Table E we find $t^* = 1.753$ for a 90% prediction interval. The interval is $\hat{y} \pm t^* s_{\hat{y}} = 136.24 \pm 1.753(8.0772) = (122.08, 150.39)$.

9.16 (a) $\bar{x} = 13.07$. $\sum(x_i - \bar{x})^2 = 443.20$. **(b)** First we calculate $s_{b_1} = s/\sqrt{\sum(x_i - \bar{x})^2} = 1.757/\sqrt{443.20} = .08346$. The null hypothesis $H_0 : \beta_1 = 0$ versus $H_a : \beta_1 > 0$ is tested by the statistic $t = b_1/s_{b_1} = .902/.08346 = 10.81$ From Table E with 8 degrees of freedom we find $P < .0005$. We conclude that the slope is positive. **(c)** The degrees of freedom are $n - 2 = 8$. From Table E we find $t^* = 3.355$ for a 99% confidence interval. The interval is $b_1 \pm t^* s_{b_1} = .902 \pm 3.355(.08346) = (.622, 1.182)$. **(d)** The estimate of the response is $\hat{y} = b_0 + b_1(15) = 1.031 + .902(15) = 14.561$. The standard error is

$$\begin{aligned} s_{\hat{y}} &= s\sqrt{1 + (1/n) + ((x^* - \bar{x})^2/\sum(x_i - \bar{x})^2)} \\ &= 1.757\sqrt{1 + (1/10) + ((15 - 13.07)^2/443.201)} \\ &= 1.8498 \end{aligned}$$

The degrees of freedom are $n - 2 = 8$. From Table E we find $t^* = 1.860$ for a 90% prediction interval. The interval is $\hat{y} \pm t^* s_{\hat{y}} = 14.561 \pm 1.860(1.8498) = (11.12, 18.00)$.

9.17 (a) $\bar{x} = 51.04$. $\sum(x_i - \bar{x})^2 = 7836.9625$. **(b)** The hypotheses are $H_0 : \beta_1 = 0$ and $H_a : \beta_1 \neq 0$. First we calculate $s_{b_1} = s/\sqrt{\sum(x_i - \bar{x})^2} = 4.4792/\sqrt{7836.9625} = .050597$. $t = b_1/s_{b_1} = 1.0935/.050597 = 21.62$ From Table E with 24 degrees of freedom we find $P < .001$. There is a relationship between the city and rural readings. **(c)** The estimated city reading is $\hat{y} = b_0 + b_1(43) = -2.580 + 1.0935(43) = 44.44$. The standard error is

$$\begin{aligned} s_{\hat{y}} &= s\sqrt{1 + (1/n) + ((x^* - \bar{x})^2/\sum(x_i - \bar{x})^2)} \\ &= 4.4792\sqrt{1 + (1/26) + ((43 - 51.04)^2/7836.9625)} \\ &= 4.58 \end{aligned}$$

The degrees of freedom are $n - 2 = 24$. From Table E we find $t^* = 2.064$ for a 95% prediction interval. The interval is $\hat{y} \pm t^* s_{\hat{y}} = 44.44 \pm 2.064(4.58) =$

$44.44 \pm 9.46 = (34.98, 53.90)$.

9.18 (a) No, there are no outliers or unusual points.

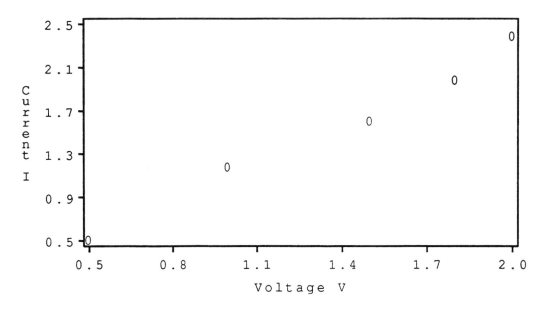

(b) The least squares fit to the data is $\hat{y} = -.06485 + 1.1844x$. The estimate for $1/R$ is the slope, 1.1844. The degrees of freedom are $n - 2 = 3$. From Table E we find $t^* = 3.182$ for a 95% confidence. The confidence interval for $1/R$ is $b_1 \pm t^* s_{b_1} = 1.1844 \pm 3.182(.07790) = (.937, 1.432)$. **(c)** The estimate of R is $\hat{R} = 1/b_1 = 1/1.1844 = .844$. The lower limit for R is $1/1.432 = .698$. The upper limit is $1/.9375 = 1.068$. The confidence interval is $(.698, 1.068)$. **(d)** The hypotheses are $H_0 : \beta_0 = 0$ versus $H_a : \beta_0 \neq 0$. The test statistic is $t = b_0/s_{b_0} = -.0649/.1142 = -.568$. The degrees of freedom are 3 and $P = .61$. The null hypothesis that the intercept is zero is not rejected.

9.19 (a) No, there are no outliers or unusual points.

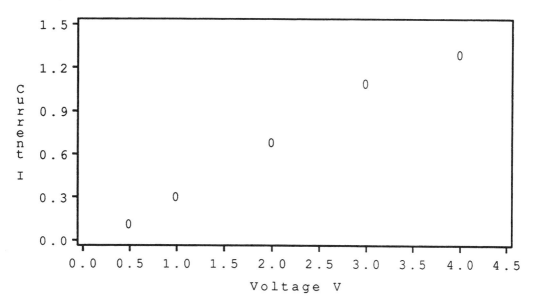

(b) The least squares fit to the data is $\hat{y} = -.027927 + .3485x$. The estimate for $1/R$ is the slope, .3485. The degrees of freedom are $n - 2 = 3$. From Table E we find $t^* = 3.182$ for a 95% confidence. The confidence interval for $1/R$ is $b_1 \pm t^* s_{b_1} = .3485 \pm 3.182(.02245) = (.2771, .4199)$. (c) The estimate of R is $\hat{R} = 1/b_1 = 1/.3485 = 2.87$. The lower limit for R is $1/.4199 = 2.38$. The upper limit is $1/.2771 = 3.61$. The confidence interval is $(2.38, 3.61)$.
(d) The hypotheses are $H_0 : \beta_0 = 0$ versus $H_a : \beta_0 \neq 0$. The test statistic is $t = b_0/s_{b_0} = -.0279/.0552 = -.506$. The degrees of freedom are 3 and $P = .65$. The null hypothesis that the intercept is zero is not rejected.

9.20 The estimate of R is $\hat{R} = 1/b_1 = 1/1.1434 = .875$. The standard error of b_1 is .02646. For a 95% confidence interval with three degrees of freedom we use $t^* = 2.776$. The confidence interval for the slope is $b_1 \pm t^* s_{b_1} = 1.1434 \pm 2.776(.02646) = (1.0699, 1.2168)$. Therefore, the confidence interval for $1/R$ is $(1/1.2168, 1/1.0699) = (.822, .935)$.

9.21 The estimate of R is $\hat{R} = 1/b_1 = 1/.3388 = 2.95$. The standard error of b_1 is .01055. For a 95% confidence interval with three degrees of freedom we use $t^* = 2.776$. The confidence interval for the slope is $b_1 \pm t^* s_{b_1} = .3388 \pm 2.776(.01055) = (.3096, .3681)$. Therefore, the confidence interval for $1/R$ is $(1/.3681, 1/.3096) = (2.7, 3.2)$.

9.22 (a) No, there are no outliers or unusual points.

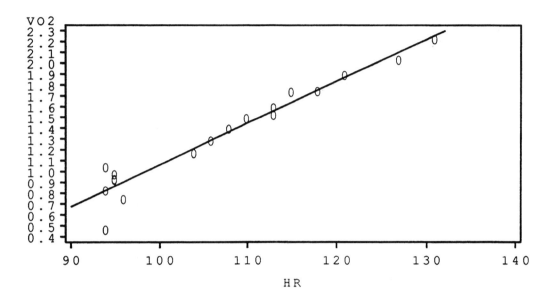

(b) $\hat{y} = -2.804 + .039x$. **(c)** $t = 16.10$, df=17, $P < .0005$, for the one-sided alternative. There is a positive linear relationship between heart rate and oxygen consumption. **(d)** The 95% prediction interval for a heart rate of 95 is (.600, 1.135), The 95% prediction interval for a heart rate of 110 is (1.186, 1.709). **(e)** Yes. The line fits the data very well and can be used in subsequent experiments.

9.23 (a) No, there are no outliers or unusual points.

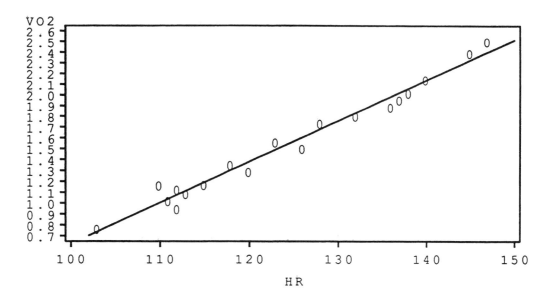

(b) $\hat{y} = -3.151 + .03777x$. **(c)** $t = 26.73$, df=17, $P < .0001$, for the one-sided alternative. There is a positive linear relationship between heart rate and oxygen consumption. **(d)** The 95% prediction interval for a heart rate of 95 is (.2442, .6306), The 95% prediction interval for a heart rate of 110 is (.8266, 1.1814). **(e)** Yes. The line fits the data very well and can be used in subsequent experiments.

9.24 The first two observations have large negative residuals. In this experiment, the apparatus was connected and no measurements were taken during a warm-up period. The data suggests that these first two observations should be considered part of the warm-up period. The regression should be rerun without these two points.

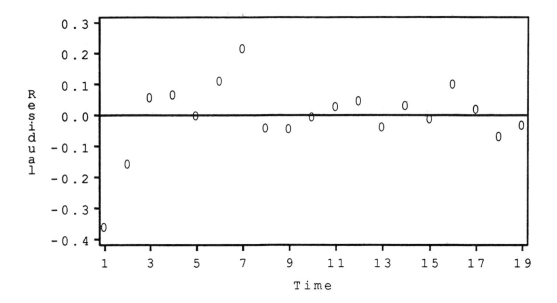

9.25 Observations two and four are rather large. There appears to be a pattern in the last six residuals. The regression methodology may be questionable in this case. More information about the data is needed.

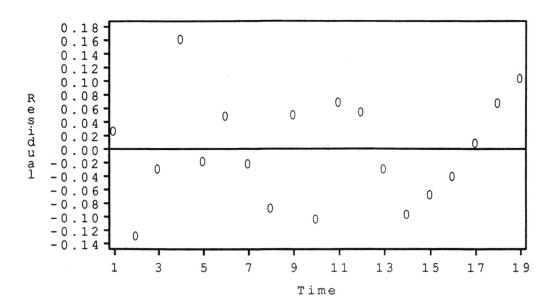

9.26 (a) The t statistic is $t = b_1/s_{b_1} = .83/.065 = 12.77$. An appropriate alternative hypothesis is one-sided, $H_a : \beta_1 > 0$. The degrees of freedom are 79 and $P < .0005$. The direct intra–arterial measure can be predicted by the oscillometric measure using a linear regression equation. (b) The predicted intra–arterial measure is $\hat{y} = 15 + 0.83(130) = 122.9$. It is not possible to compute a prediction interval because the information needed for the standard error is not given in the problem. We can compute a 99% confidence interval for the slope as follows. From Table E we find $t^* = 2.639$ for a 99% confidence interval with 80 degrees of freedom. $b_1 \pm t^* s_{b_1} = .83 \pm 2.639(.065) = (.658, 1.002)$.

9.27 (a) The t statistic is $t = b_1/s_{b_1} = .00665/.00182 = 3.65$. An appropriate alternative hypothesis is one-sided, $H_a : \beta_1 > 0$. The degrees of freedom are 16 and $.0010 \leq P \leq .0025$. We conclude there is a positive linear relation between air flow and evaporation. (b) The additional evaporation experienced when air speed increases by 1 unit is the slope of the true regression line β_1. From Table E we find $t^* = 2.120$ for a 95% confidence interval. The confidence interval is $b_1 \pm t^* s_{b_1} = .00665 \pm 2.120(.00182) = (.0028, .0105)$.

9.28 (a)

Source	df	SS	MS	F	P
Model	1	2.093	2.093	231.21	.0006
Error	3	.027	.009		
Total	4	2.120			

(b) $H_0 : \beta_1 = 0$. The null hypothesis is that there is no linear relation between current and voltage. (c) $F(1, 3)$, $P < .001$.

9.29 (a)

Source	df	SS	MS	F	P
Model	1	.9961	.9961	240.95	.0006
Error	3	.0124	.0041		
Total	4	1.0085			

(b) $H_0 : \beta_1 = 0$. The null hypothesis is that there is no linear relation between current and voltage. (c) $F(1, 3)$, $P < .001$.

9.30 (a)

Source	df	SS	MS	F	P
Model	1	3.762	3.762	259.27	.0001
Error	17	.247	.015		
Total	18	4.009			

(b) $H_0 : \beta_1 = 0$. This hypothesis states that there is no linear relation between heart rate and oxygen consumption. (c) $F(1,17)$, $P < .001$. (d) $t^2 = (16.102)^2 = 259.27$. (e) $R^2 = .94$.

9.31 (a)

Source	df	SS	MS	F	P
Model	1	4.5125	4.5125	714.618	.0001
Error	17	.1073	.0063		
Total	18	4.6198			

(b) $H_0 : \beta_1 = 0$. This hypothesis states that there is no linear relation between heart rate and oxygen consumption. (c) $F(1,17)$, $P < .001$. (d) $t^2 = (26.73)^2 = 714.493$. The difference is due to roundoff. (e) $R^2 = .9768$.

9.32 (a) $t = r\sqrt{n-2}/\sqrt{1-r^2} = .39\sqrt{40-2}/\sqrt{1-.39^2} = 2.61$. (b) $H_a : \rho > 0$. (c) Using Table E with 40 degrees of freedom we find $.005 \leq P \leq .01$. The exact value is $P = .006$. We conclude that higher birth rates are associated with higher incomes.

9.33 (a) We test $H_0 : \rho = 0$ versus $H_a : \rho > 0$. The test statistic is $t = r\sqrt{n-2}/\sqrt{1-r^2} = -.19\sqrt{713-2}/\sqrt{1-(-.19)^2} = -5.16$. (b) Using Table E with 100 or 1000 degrees of freedom we find $P \leq .001$. The exact value is $P = .0000002$. We conclude that there is a positive linear association between parental control and self–esteem of the students. Note that the result of the statistical test is highly significant but the correlation is rather small. This is partly a consequence of the large sample size.

9.34 (a) $\mu_{GPA} = \beta_0 + 9\beta_1 + 8\beta_2 + 7\beta_3$. (b) The estimate of the subpopulation mean is $\hat{\mu} = 2.5899 + 9(.16857) + 8(.034316) + 7(.045102) = 4.697$.

9.35 (a) $\mu_{GPA} = \beta_0 + 6\beta_1 + 7\beta_2 + 8\beta_3$. **(b)** The estimate of the subpopulation mean is $\hat{\mu} = 2.5899 + 6(.16857) + 7(.034316) + 8(.045102) = 4.202$. This is an estimate of the mean GPA for all students who have high school grades of B– in math, B in science, and B+ in English.

9.36 (a) The degrees of freedom are 220. Using Table E we find $t^* = 1.962$ for 1000 degrees of freedom. The exact value is $t^* = 1.971$. The confidence interval for the coefficient of HSM is $b \pm t^* s_b = .16857 \pm 1.971(.03549) = (.099, .239)$. The regression coefficient of HSM is the change in GPA that is associated with an increase of one point in high school math grades given that the other two high school grades remain the same. **(b)** The confidence interval for the coefficient of HSE is $b \pm t^* s_b = .045102 \pm 1.971(.03870) = (-.031, .121)$. The regression coefficient of HSE is the change in GPA that is associated with an increase of one point in high school math grades given that the other two high school grades remain the same.

9.37 (a) The degrees of freedom are 221. Using Table E we find $t^* = 1.962$ for 1000 degrees of freedom. The exact value is $t^* = 1.971$. The confidence interval for the coefficient of HSM is $b \pm t^* s_b = .18265 \pm 1.971(.031956) = (.120, .246)$. Assuming that the other high school grades remain constant, an increase of one unit in high school math grades is associated with an increase of $\beta_1 = .183$ units in GPA. **(b)** The confidence interval for the coefficient of HSE is $b \pm t^* s_b = .06067 \pm 1.971(.03473) = (-.008, .129)$. Assuming that the other high school grades remain constant, an increase of one unit in high school English grades is associated with an increase of $\beta_3 = .061$ units in GPA. The confidence interval indicates that this increase is not distinguishable from zero. The results are different from those given in the previous exercise because only two explanatory variables are used in this exercise, whereas three were used in the previous exercise. The regression coefficient for a given variable depends upon the other variables that are present in the model.

9.38 (a) $\hat{y} = 2.590 + .169x_1 + .034x_2 + .045x_3$. **(b)** The estimate of σ is $s = .6998$. This value is given in Figure 9.21 as ROOT MSE. **(c)** $H_0 : \beta_1 = \beta_2 = \beta_3 = 0$, $H_a : \beta_j \neq 0$ for at least one $j=1,2,3$. The null hypothesis states that none of the three high school grades have predictive value in a linear regression model. The alternative states that at least one of the three high school grades is linearly related to GPA. (d) $F(3, 220)$. We conclude that at least one of the three high school grades can be used in a linear model to predict GPA ($P = .0001$). (e) $R^2 = 20.46\%$.

9.39 (a) $\hat{y} = 3.2887 + .002283x_1 - .00002456x_2$. (b) The estimate of σ is $s = .7577$. This value is given in Figure 9.24 as ROOT MSE. (c) $H_0 : \beta_1 = \beta_2 = 0$, $H_a : \beta_j \neq 0$ for at least one $j=1,2$. The null hypothesis states that neither of the two SAT scores have predictive value in a linear regression model. The alternative states that at least one of the two SAT scores is linearly related to GPA. (d) $F(2, 221)$. We conclude that at least one of the two SAT scores can be used in a linear model to predict GPA ($P = .0007$). (e) $R^2 = 6.34\%$.

9.40 A 95% prediction interval is $2.136 \pm 1.984(.013) = 2.136 \pm .026 = (2.110, 2.162)$. For t^* we use the value corresponding to 100 degrees of freedom for a 95% interval from Table E. The predicted value from the regression equation is very close to the observed value. The prediction interval includes the observed value. Therefore, the actual price is very compatible with the rest of the available data and there is no evidence to conclude that the price was artificially depressed. We agree with the judge.

9.41 For SATM and GPA a hint of a positive relationship can be seen. This corresponds to the statistically significant ($P=.0001$) correlation of .251 reported in Figure 9.20. No clear relationship is evident between SATV and GPA. There are no outliers or unusual points.

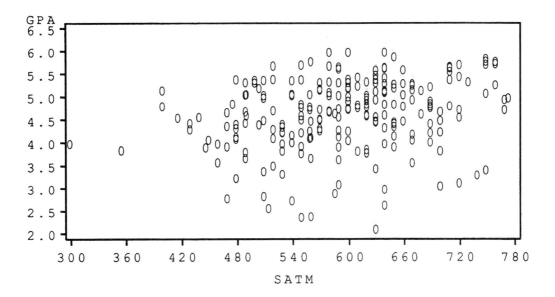

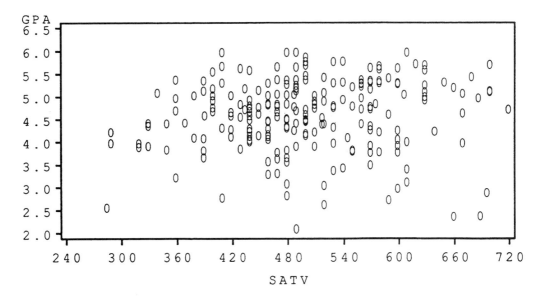

9.42 It is difficult to see relationships in these plots. They suggest some rather small positive relationships. No outliers or unusual observations are evident. The discreteness of the data is evident. With a plot from a line printer, many observations are plotted in the same position. By adding a small random amount to each observation a better graph can be made with a plotter or graphics device.

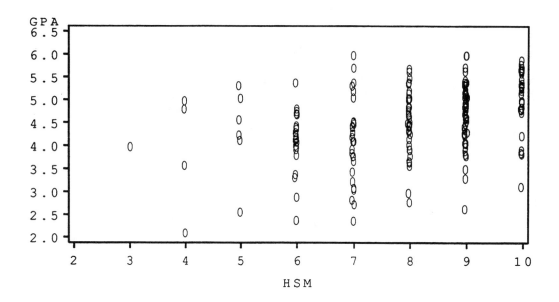
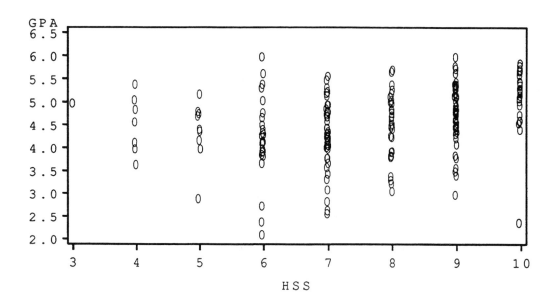

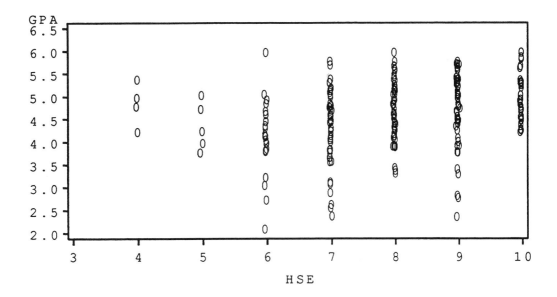

9.43 No unusual points or patterns are evident in the residual plots.

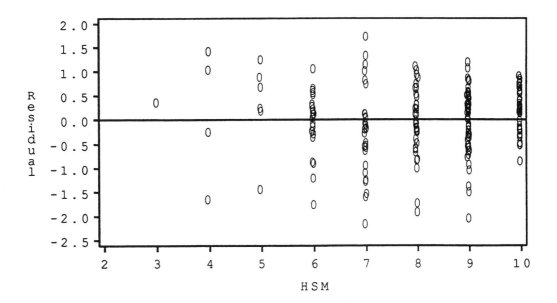

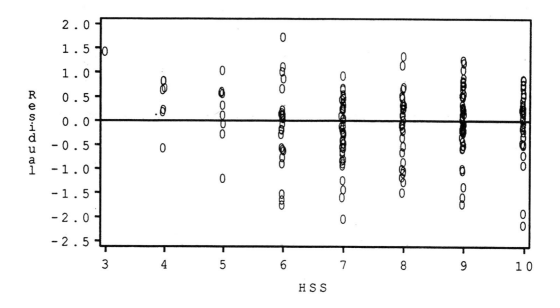

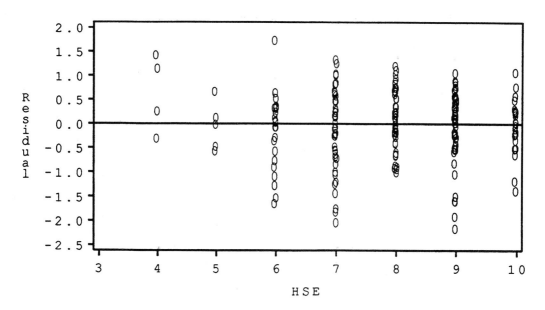

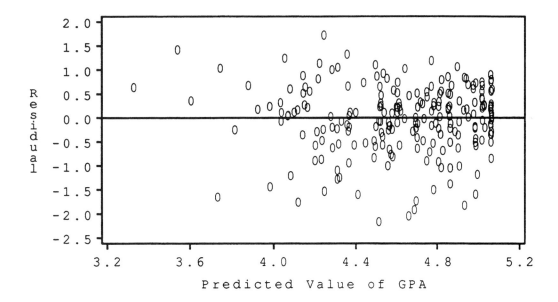

9.44 There appears to be some skewness in the distribution of the residuals. The lower tail stretches out more than the upper tail. No other clear patterns are evident.

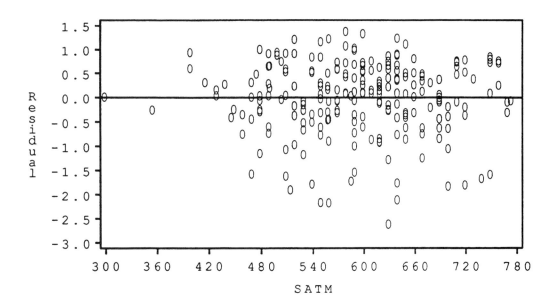

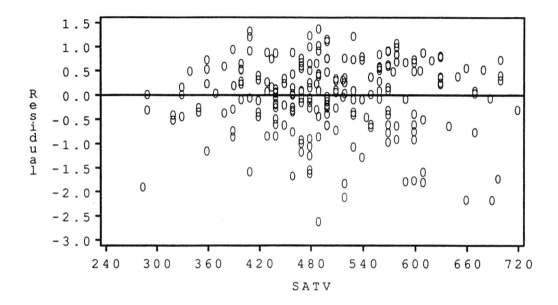

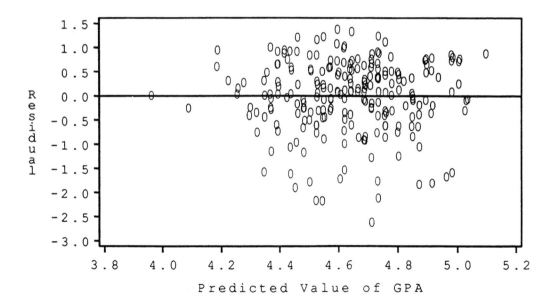

9.45 (a) $\hat{y} = 2.666 + .193HSM + .0006SATM$. **(b)** $H_0 : \beta_1 = \beta_2 = 0$, $H_a : \beta_j \neq 0$ for at least one $j=1,2$. The null hypothesis states that neither high school math grades nor SAT math score predicts GPA in a linear regression model. The alternative states that at least one of these has predictive value in a regression model with GPA as the response variable. $F=26.63$.

$P < .0001$. The null hypothesis is rejected and we conclude that at least one of HSM and SATM has a linear relation with GPA. **(c)** The degrees of freedom are $n - 2 = 221$. From Table E we find $t^* = 1.984$ for 100 degrees of freedom and $t^* = 1.962$ for 1000 degrees of freedom. The exact value is $t^* = 1.971$. The 95% confidence interval for the regression coefficient of HSM is $b_1 \pm t^* s_{b_1} = .1930 \pm 1.971(.03222) = (.1295, .2565)$. The 95% confidence interval for the regression coefficient of SATM is $b_2 \pm t^* s_{b_2} = .0006105 \pm 1.971(.0006112) = (-.0006, .0018)$. The second interval includes 0. **(d)** For HSM, the test statistic is $t = 5.99$ and $P < .001$. For SATM, the test statistic is $t = 1.00$ and $P = .319$. For this model we reject $H_0 : \beta_1 = 0$ and we do not reject $H_0 : \beta_2 = 0$. Given that HSM is in the model, SATM does add statistically significant information for predicting GPA. **(e)** $s = .7028$. **(f)** $R^2 = 19.42\%$.

9.46 The regression equation is $\hat{y} = 3.2750 + .1435 HSE + .0003942 SATV$. The hypotheses are $H_0 : \beta_1 = \beta_2 = 0$, $H_a : \beta_j \neq 0$ for at least one $j=1,2$. $F=10.34$. $P < .0001$. We reject the null hypothesis. The degrees of freedom are $n - 2 = 221$. From Table E we find $t^* = 1.984$ for 100 degrees of freedom and $t^* = 1.962$ for 1000 degrees of freedom. The exact value is $t^* = 1.971$. The 95% confidence interval for the regression coefficient of HSE is $b_1 \pm t^* s_{b_1} = .1435 \pm 1.971(.0343) = (.0759, .2111)$. The 95% confidence interval for the regression coefficient of SATV is $b_2 \pm t^* s_{b_2} = .0003942 \pm 1.971(.000558) = (-.0007, .0015)$. The second interval includes 0. For HSE, the test statistic is $t = 4.185$ and $P < .0001$. For SATV, the test statistic is $t = .706$ and $P = .481$. For this model we reject $H_0 : \beta_1 = 0$ and we do not reject $H_0 : \beta_2 = 0$. Given that HSE is in the model, SATV does add statistically significant information for predicting GPA. $s = .7487$. $R^2 = 8.56\%$. This model gives a much lower R^2. This means that the math variables are better predictors than the verbal variables. In both models the high school grades predict GPA better than the SAT scores.

9.47 (a) There appears to be a strong positive linear relationship between the two measurements. There are no outliers or unusual points.

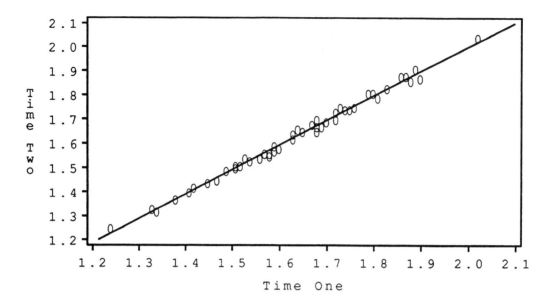

(b) $\widehat{T2} = .0048 + .9920 T1$, $s = .0265$. (c) $r = .988$, $r^2 = .977$. (d) $t = 45.22$, $H_a : \beta_1 > 0$, $P < .0001$. We conclude that there is a very strong relationship between the two measurements. (e) $t^2 = (45.22)^2 = 2044.85$, $F = 2044.66$. Difference is due to roundoff.

9.48 (a)

	n	b_0	b_1	s	s_{b_1}
Run 10.41	50	.004817	.9920	.02648	.02194
Run 10.42	25	-.0581	1.0311	.01527	.01701

Relative to the standard errors the regression coefficients have not changed. The standard errors have decreased. (b) The ANOVA table for the full data set is

Source	df	SS	MS	F	P
Model	1	1.433	1.433	2044.66	.0001
Error	48	.03365	.00070		
Total	49	1.467			

The ANOVA table for the odd numbered strips is

Source	df	SS	MS	F	P
Model	1	.8568	.8568	3673.71	.0001
Error	23	.0054	.00023		
Total	24	.8622			

All sums of squares and mean squares are smaller for the reduced sample. The relationship is so strong that no difference in the significance is evident. (c) The correlation in Exercise 9.41 is .9885. For this exercise it is .9969. The two values are quite close. (d) We would expect weaker results with the smaller sample size. For this data set the results are approximately the same and the F value is actually larger for the subset of the data. (e) With a larger sample we would expect the values of b_0, b_1, s and r to be approximately the same as those obtained for the sample of size 50.

9.49 (a) $\hat{y} = 109.87 - 1.127x$. (b) For testing the null hypothesis that the slope is zero we use the statistic $t = -3.63$. It has the $t(19)$ distribution. The P-value for a one-sided alternative is $P = .0009$. (c) The degrees of freedom are $n - 2 = 19$. From Table E we find $t^* = 2.093$ for a 95% confidence interval. $b_1 \pm t^* s_{b_1} = -1.127 \pm 2.093(.3102) = (-1.776, -.478)$. (d) $R^2 = 41.00\%$. (e) The estimate of the model standard deviation, σ, is $s = 11.023$.

9.50 (a) $\hat{y} = 107.59 - 1.050x$. (b) For testing the null hypothesis that the slope is zero we use the statistic $t = -2.508$. It has the $t(17)$ distribution. The P-value for a one-sided alternative is $P = .0113$. (c) The degrees of freedom are $n - 2 = 17$. From Table E we find $t^* = 2.110$ for a 95% confidence interval. $b_1 \pm t^* s_{b_1} = -1.050 \pm 2.110(.4186) = (-1.933, -.167)$. (d) $R^2 = 27.01\%$. (e) The estimate of the model standard deviation, σ, is $s = 8.831$. Deletion of the influential observation has caused the relationship to be less strong although it is still statistically significant. The removal of the outlier reduces the value of s.

9.51 Plots of GPA versus each predictor do not indicate any very strong relationships. The F statistic for the model is 9.63 with 3 and 141 degrees of freedom and $P < .0001$. The proportion of variation explained is .1844. The model with the three high school variables as predictors is statistically significant. For the tests on the individual regression coefficients, only HSM is significant ($t = 3.46$, df=141, $P = .0007$). No unusual patterns or observations are evident in the residual plots. The model should be rerun deleting either HSS or HSE. The results for males alone are similar to those obtained for all students in the case study. The value of R^2 is a little smaller (.1844 for males versus .2046 for all students). The overall model and the HSM coefficient are slightly less significant, partly because of the reduced sample size.

9.52 Plots of GPA versus each predictor do not indicate any very strong relationships. The F statistic for the model is 8.385 with 3 and 75 degrees of freedom and $P < .0001$. The proportion of variation explained is .2512. The model with the three high school variables as predictors is statistically significant. For the tests on the individual regression coefficients, only HSM is significant ($t = 3.344$, df=75, $P = .0013$). No unusual patterns or observations are evident in the residual plots. The model should be rerun deleting either HSS or HSE. The results for females are very similar to the results for males. The R^2 values are .2512 for females and .1844 for males. Both regression models are statistically significant and HSM appears to be the only important predictor.

9.53 (a) $\hat{y} = -9.102 + 1.086x$. (b) The standard error of the slope is $s_{b_1} = .6529$. (c) The test statistic is $t = 1.66$ with 7 degrees of freedom. $P = .14$. All of the results are quite different. The intercept is now negative. The value of t has changed from 17.64 (see Example 9.5) to 1.66 and is no longer statistically significant. The incorrect observation has caused the results of the regression analysis to be meaningless.

9.54

	$\bar{x}$	Median	s	IQR
Taste	24.533	20.95	16.255	24.575
Acetic	5.498	5.425	.571	.713
H2S	5.942	5.329	2.127	3.766
Lactic	1.442	1.45	.303	.463

There are three rather high values for taste and the distribution appears somewhat skewed toward high values. The distribution of acetic is approximately normal. H2S is skewed toward high values. Lactic is approximately normal. The following stemplots were constructed using rounded values. Note that in the plot for acetic, pairs of stems are combined.

```
Taste                       Acetic
  0 |  11                     44 |  846
  0 |  566                    46 |  69
  1 |  2234                   48 |  0
  1 |  56788                  50 |  6
  2 |  112                    52 |  4450377
  2 |  666                    54 |  145
  3 |  2                      56 |  046
  3 |  5799                   58 |  059
  4 |  1                      60 |  4858
  4 |  8                      62 |  6
  5 |                         64 |  56
  5 |  577

H2S                         Lactic
  3 |  01278899                8 |  6
  4 |  27899                   9 |  9
  5 |  024                    10 |  689
  6 |  1278                   11 |  56
  7 |  0569                   12 |  5599
  8 |  07                     13 |  013
  9 |  126                    14 |  469
 10 |  2                      15 |  2378
                              16 |  38
                              17 |  248
                              18 |  1
                              19 |  09
                              20 |  1
```

The normal quantile plots are

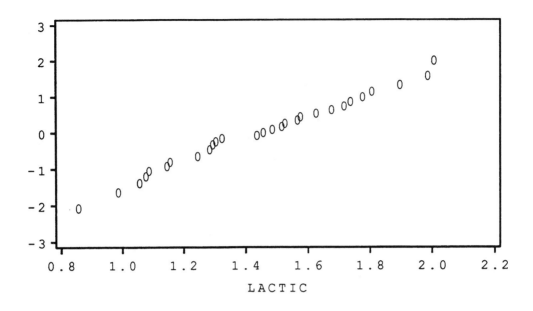

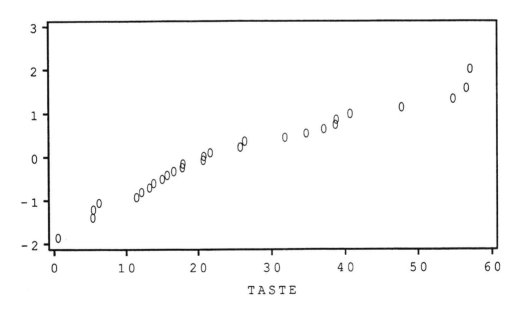

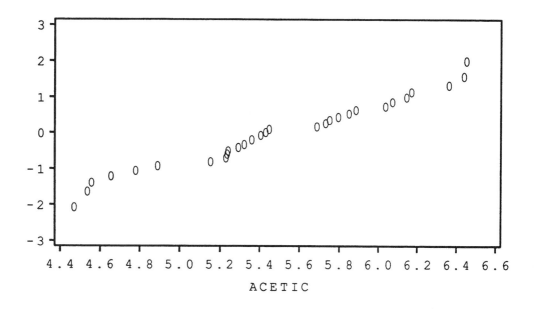

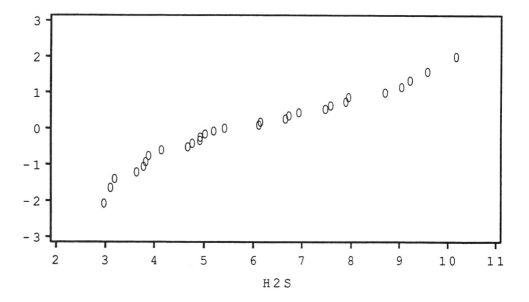

9.55 The relationships are weak and positive. The correlations and their P-values are: taste with acetic .55, .0017; taste with H2S .76, .0001; taste with lactic .70, .0001; acetic with H2S .62, .0003; acetic with lactic .60, .0004; H2S with lactic .64, .0001.

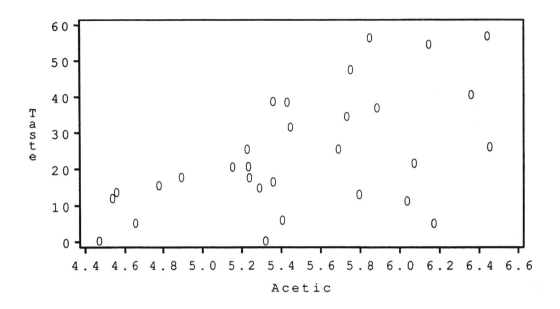
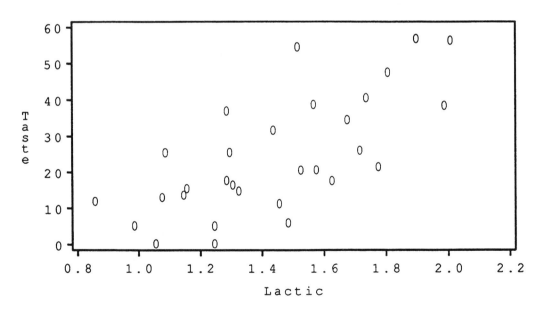

CHAPTER 9

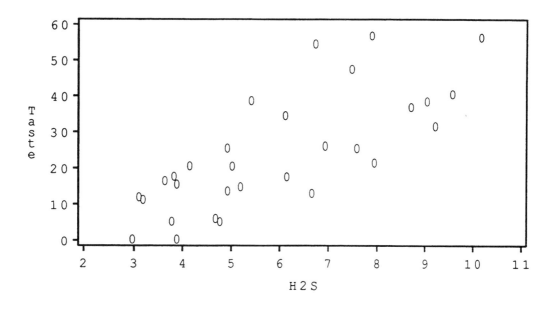

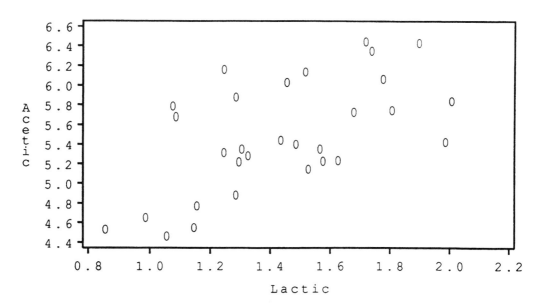

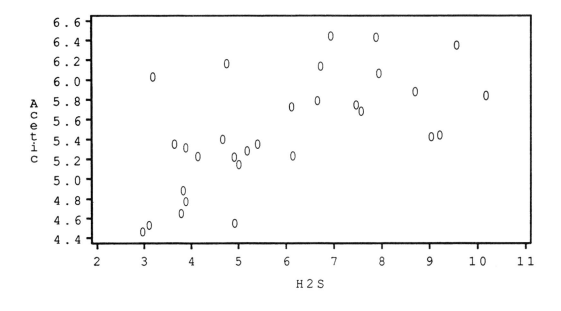

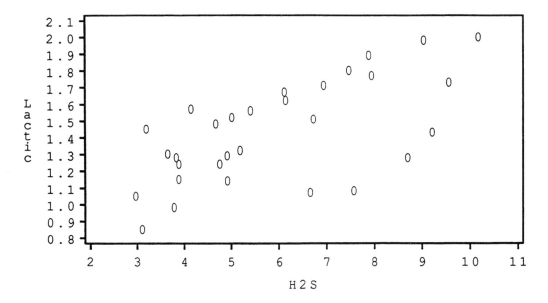

9.56 The fitted equation is $\hat{y} = -61.50 + 15.65x$. For the model, $F = 12.114$ with 1 and 28 degrees of freedom. The P-value is .0017 and $R^2 = .30$. The residuals appear to be somewhat smaller for low values of acetic. For large values of H2S and lactic the residuals are all positive. This suggests that the model could be improved by adding one or more of these variables.

CHAPTER 9

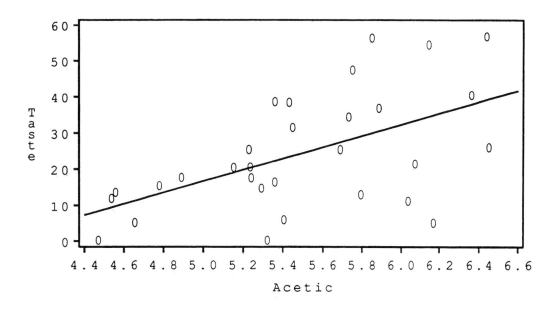

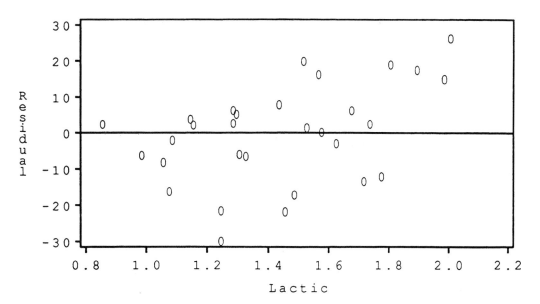

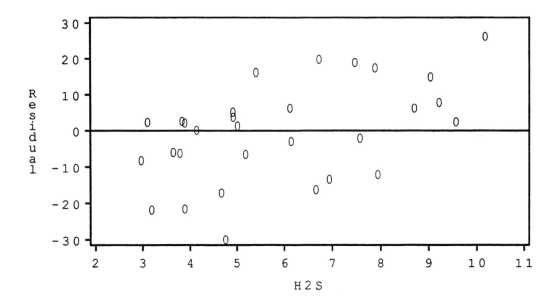

9.57 The fitted equation is $\hat{y} = -9.79 + 5.78x$. For the model, $F = 37.29$ with 1 and 28 degrees of freedom. The P-value is .0001 and $R^2 = .57$. The residuals are smaller for low values of H2S. The residual plots versus acetic and lactic show no clear patterns. There is some tendency for smaller residuals to be associated with smaller values of these variables. The normal quantile plot of the residuals indicates that they are approximately normal with a small amount of skewness to the right.

CHAPTER 9

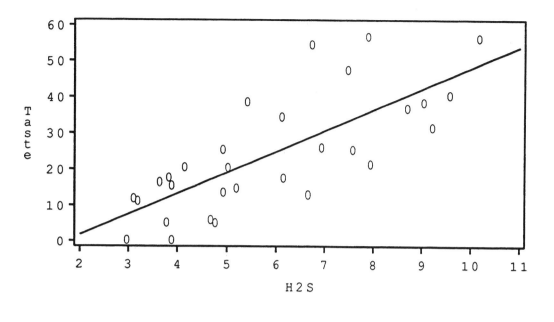

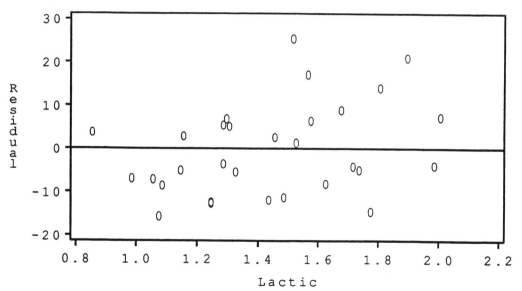

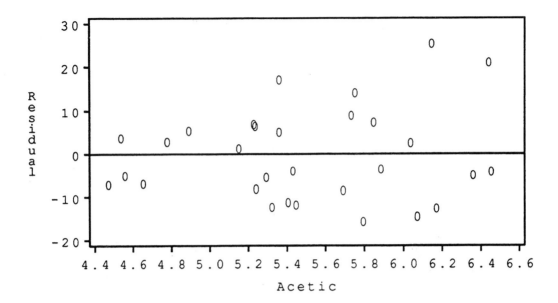

9.58 The fitted equation is $\hat{y} = -29.86 + 37.72$. For the model, $F = 27.55$ with 1 and 28 degrees of freedom. The P-value is .0001 and $R^2 = .50$. The residuals appear to more variable for the higher values of acetic. A weak linear relationship between the residuals and H2S is evident. Addition of this variable to the model may improve the fit. No clear pattern in the plot of the residuals versus lactic. The normal quantile plot of the residuals indicates that they are approximately normal with a small amount of skewness to the right.

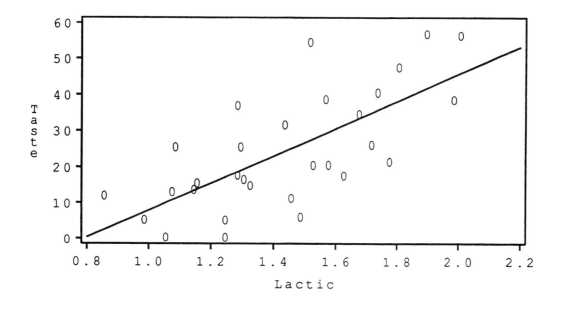
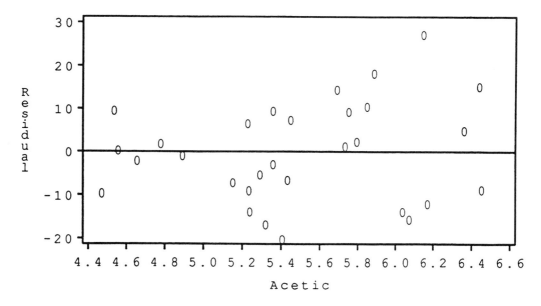

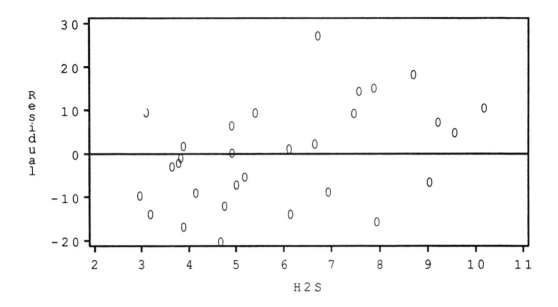

9.59

	F	P	R^2	$\sqrt{MSE}$	b_0	b_1
Acetic	12.114	.0017	.30	13.82	-61.50	15.65
H2S	37.293	.0001	.57	10.83	-9.79	5.78
Lactic	27.55	.0001	.50	11.75	-29.86	37.72

The regression equations are

$$\hat{y} = -61.50 + 15.65 Acetic$$
$$\hat{y} = -9.79 + 5.78 H2S$$
$$\hat{y} = -29.86 + 37.72 Lactic$$

The intercepts are not the same because the models are different.

9.60 The fitted equation is $\hat{y} = -26.94 + 3.80 Acetic + 5.15 H2S$. For the model, $F = 18.81$ with 2 and 27 degrees of freedom. The P-value is .0001 and $R^2 = .58$. The coefficient for acetic is not statistically significant ($t = .844$, $df = 27$, $P = .41$), while the coefficient for H2S is clearly significant ($t = 4.255$, $df = 27$, $P = .0002$). For the model with acetic as the only explanatory variable it is clearly significant. However, when H2S is added to the model, it is no longer significant. This analysis suggests that the information in acetic that is useful for predicting taste, is also present in H2S. The model with both variables predicts taste better than a model with acetic alone ($R^2 = 58\%$ versus 30%).

9.61 For the multiple regression $\hat{y} = -27.59 + 3.95H2S + 19.89Lactic$. The coefficient for H2S is significant ($t = 3.475$, $P = .0018$). The same is true for the coefficient for Lactic ($t = 2.499$, $P = .0188$). For this model $F = 25.26$, $P = .0001$, $R^2 \doteq .652$. The R^2s for the simple regressions are .5712 for H2S and .4959 for Lactic. We prefer the multiple regression because both coefficients are significant and the R^2 is considerably higher.

9.62 For the multiple regression $\hat{y} = -28.88 + .33Acetic + 3.91H2S + 19.67Lactic$. The coefficient for acetic is not significant ($t = .073$, $df = 26$, $P = .9420$). The coefficient for H2S is significant ($t = 3.133$, $df = 26$, $P = .0043$). The same is true for the coefficient for Lactic ($t = 2.280$, $df = 26$, $P = .0311$). For this model $F = 16.221$ with 3 and 26 degrees of freedom. The P-value is .0001 and $R^2 = .652$. The residuals show no clear patterns except for a tendency toward larger variation for high values of the explanatory variables. The model with three explanatory variables has essentially the same R^2 as the model with H2S and lactic (.6517 versus .6518). Note that acetic is not significant in the three variable model. The models with only one explanatory variable do not fit as well. Therefore, we prefer the model with H2S and lactic as explanatory variables.

4.10 CHAPTER 10

10.1 (a) The response variable is yield of tomatoes in pounds and the populations to be compared are varieties of tomatoes. $I = 4$; $n_i = 10, 10, 10, 10$; and $N = 40$. **(b)** The response variable is attractiveness rating and the populations to be compared are types of packaging for the laundry detergent. $I = 6$; $n_i = 120, 120, 120, 120, 120, 120$; and $N = 720$. **(c)** The response variable is weight loss and the populations to be compared are the types of weight loss programs. $I = 3$; $n_i = 10, 10, 10$; and $N = 30$.

10.2 (a) The response variable is the number of hours of sleep on a typical night and the populations to be compared are the groups of people classified by smoking behavior. $I = 3$; $n_i = 200, 200, 200$; and $N = 600$. **(b)** The response variable is the strength of the cement and the populations to be compared are the types of cement mixture. $I = 5$; $n_i = 6, 6, 6, 6, 6$; and $N = 30$. **(c)** The response variable is the score on a final exam and the populations to be compared are students taught by the different methods. $I = 4$; $n_i = 10, 10, 10, 10$; and $N = 40$.

10.3 Yes we can pool because the ratio of the largest standard deviation to the smallest standard deviation is $10.1/5.2 = 1.94 < 2$. The pooled variance is $(19(5.2)^2 + 19(8.9)^2 + 19(10.1)^2)/(19+19+19) = 69.42$. $s_p = \sqrt{69.42} = 8.33$.

10.4 Yes we can pool because the ratio of the largest standard deviation to the smallest standard deviation is $12.2/9.2 = 1.33 < 2$. The pooled variance is $(91(12.2)^2 + 33(10.4)^2 + 34(9.2)^2 + 23(11.7)^2)/(91+33+34+23) = 127.85$. $s_p = \sqrt{127.85} = 11.31$.

10.5 (a)

Source	df
Variety	3
Error	36
Total	39

(b)

Source	df
Type	5
Error	714
Total	719

(c)

Source	df
Type	2
Error	27
Total	29

10.6 (a)

Source	df
Groups	2
Error	597
Total	599

(b)

Source	df
Groups	4
Error	25
Total	29

(c)

Source	df
Groups	3
Error	36
Total	39

10.7 (a) The hypotheses are $H_0 : \mu_1 = \mu_2 = \mu_3$ and H_a : Not all of the μ_i are equal. **(b)**

Source	df
Major	2
Error	253
Total	255

(c) $F(2, 253)$. (d) Using Table F, the critical value is 3.04 for $F(2, 200)$; 3.00 for $F(2, 1000)$. From a computer, the exact value is 3.03 for $F(2, 253)$.

10.8 (a) The hypotheses are $H_0 : \mu_1 = \mu_2 = \mu_3 = \mu_4$ and H_a : Not all of the μ_i are equal. **(b)**

Source	df
Classes	3
Error	196
Total	199

(c) $F(3, 196)$. (d) Using Table F, the critical value is 2.65 for $F(3, 200)$. From a computer, the exact value is 2.65 for $F(3, 196)$.

10.9 (a)

Source	SS	DF	MS	F
Groups	104855.87	3	34951.96	15.86
Error	70500.59	32	2203.14	
Total	175356.46	35		

(b) The hypotheses are $H_0 : \mu_1 = \mu_2 = \mu_3 = \mu_4$ and H_a : Not all of the μ_i are equal. (c) $F(3, 32)$. $P < .001$. The mean fitness scores for the four groups are not all the same. (d) $s_p^2 = $ MSE $= 2203.14$. $s_p = \sqrt{s_p^2} = \sqrt{2203.14} = 46.94$.

10.10 (a)

Source	SS	DF	MS	F
Groups	476.87	3	158.9567	2.53
Error	2009.88	32	62.8088	
Total	2486.75	35		

(b) The hypotheses are $H_0 : \mu_1 = \mu_2 = \mu_3 = \mu_4$ and H_a : Not all of the μ_i are equal. (c) $F(3, 32)$. Using Table F with 3 and 30 degrees of freedom we find $.05 \leq P \leq .10$. The same result is obtained using 3 and 40 degrees of freedom. The exact value is $P = .0746$. We do not have evidence to conclude that the groups are different in the mean depression score. (d) $s_p^2 = $ MSE $=$

62.8088. $s_p = \sqrt{s_p^2} = \sqrt{62.8088} = 7.93$.

10.11 (a) $s_p^2 = (45(2.5)^2 + 110(1.8)^2 + 51(1.8)^2)/(45 + 110 + 51) = 3.90$. We find this value as MSE in the ANOVA table. **(b)**

Source	SS	DF	MS	F
Country	17.22	2	8.61	2.21
Error	802.89	206	3.90	
Total	820.11	208		

(c) The hypotheses are $H_0 : \mu_1 = \mu_2 = \mu_3$ and H_a : Not all of the μ_i are equal. **(d)** $F(2, 206)$. Using Table F with 2 and 200 degrees of freedom we find $P > .1$. From a computer, $P = .112$. The data do not provide evidence to conclude that the mean birth weights are different in the three countries. **(e)** $R^2 = SSG/SST = 17.22/820.11 = .0210$.

10.12 (a) $s_p^2 = (87(327)^2 + 90(184)^2 + 53(285)^2)/(87 + 90 + 53) = 72412.12$. We find this value as MSE in the ANOVA table. **(b)**

Source	SS	DF	MS	F
Country	6572551	2	3286276	45.38
Error	16654788	230	72412	
Total	23227339	232		

(c) The hypotheses are $H_0 : \mu_1 = \mu_2 = \mu_3$, H_a : Not all of the μ_i are equal. **(d)** $F(2, 230)$. Using Table F with 2 and 200 degrees of freedom we find $P < .001$. We conclude that the food intakes in the three countries are different. **(e)** $R^2 = SSG/SST = 6572551/23227339 = .283$.

10.13 (a) $\psi_1 = \frac{1}{2}(\mu_1 + \mu_2) - \mu_3$. **(b)** $\psi_2 = \mu_1 - \mu_2$.

10.14 Each of the following contrasts could be expressed with all signs reversed. **(a)** $\psi_1 = -\frac{1}{2}(\mu_1 + \mu_2) + \frac{1}{2}(\mu_3 + \mu_4)$. **(b)** $\psi_2 = \mu_1 - \mu_2$. **(c)** $\psi_3 = \mu_3 - \mu_4$.

10.15 (a) The hypotheses for the first contrast are $H_{01} : \frac{1}{2}(\mu_1 + \mu_2) = \mu_3$, $H_{a1} : \frac{1}{2}(\mu_1 + \mu_2) > \mu_3$. The hypotheses for the second contrast are $H_{02} : \mu_1 = \mu_2$, $H_{a2} : \mu_1 \neq \mu_2$. **(b)** The first sample contrast is $c = \frac{1}{2}(\bar{x}_1 + \bar{x}_2) - \bar{x}_3 =$

$\frac{1}{2}(619 + 629) - 575 = 49$. The second sample contrast is $c = \bar{x}_1 - \bar{x}_2 = 619 - 629 = -10$. (c) Using the formula

$$s_c = s_p\sqrt{\sum(a_i^2/n_i)}$$

we calculate the standard error for the first sample contrast as

$$s_c = 82.5\sqrt{.5^2/103 + .5^2/31 + (-1)^2/122}$$
$$= 11.28$$

and the standard error for the second sample contrast as

$$s_c = 82.5\sqrt{1^2/103 + (-1)^2/31}$$
$$= 16.90$$

(d) The test statistic for the first contrast is $t_1 = c/s_c = 49/11.28 = 4.34$, with df=253, and $P < .0005$. The test statistic for the second contrast is $t_2 = c/s_c = -10/16.90 = -.59$, with df=253, and $P > .50$. From the computer $P = .555$. The average of the means for the first two groups is greater than the mean for the third group. There is no evidence to conclude that the first two groups have different means. (e) From Table E we find $t^* = 1.984$ for 100 degrees of freedom and $t^* = 1.962$ for 1000 degrees of freedom. The exact value is $t^* = 1.969$. The confidence intervals are $c \pm t^* s_c = 49 \pm 1.969(11.28) = (26.78, 71.22)$ and $c \pm t^* s_c = -10 \pm 1.969(16.90) = (-43.29, 23.29)$.

10.16 (a) The hypotheses for the first contrast are $H_{01} : \frac{1}{2}(\mu_1 + \mu_2) = \mu_3$, $H_{a1} : \frac{1}{2}(\mu_1 + \mu_2) > \mu_3$. The hypotheses for the second contrast are $H_{02} : \mu_1 = \mu_2$, $H_{a2} : \mu_1 \neq \mu_2$. (b) The first sample contrast is $c = \frac{1}{2}(\bar{x}_1 + \bar{x}_2) - \bar{x}_3 = \frac{1}{2}(8.77 + 8.75) - 7.83 = .93$. The second sample contrast is $c = \bar{x}_1 - \bar{x}_2 = 8.77 - 8.75 = .02$. (c) Using the formula

$$s_c = s_p\sqrt{\sum(a_i^2/n_i)}$$

we calculate the standard error for the first contrast

$$s_c = 1.158\sqrt{.5^2/90 + .5^2/28 + (-1)^2/106}$$
$$= .2299$$

and the standard error for the second contrast

$$s_c = 1.158\sqrt{1^2/90 + (-1)^2/28}$$
$$= .3421$$

(d) The test statistic for the first contrast is $t_1 = c/s_c = .93/.2299 = 4.046$, with df=221, and $P < .0005$. The test statistic for the second contrast is $t_2 = c/s_c = .02/.3421 = .0585$, with $df = 221$, and $P > .50$. From the computer $P = .953$. The average of the means for the first two groups is greater than the mean for the third group. There is no evidence to conclude that the first two groups have different means. **(e)** From Table E we find $t^* = 1.984$ for 100 degrees of freedom and $t^* = 1.962$ for 1000 degrees of freedom. The exact value is $t^* = 1.971$. The confidence intervals are $c \pm t^* s_c = .93 \pm 1.971(.2299) = (.477, 1.383)$ and $c \pm t^* s_c = .02 \pm 1.971(.3421) = (-.65, .69)$.

10.17 (a) The contrasts are $\psi_1 = \mu_1 - \mu_2$, $\psi_2 = \mu_1 - \frac{1}{2}(\mu_2 + \mu_4)$, and $\psi_3 = \mu_3 - \frac{1}{3}(\mu_1 + \mu_2 + \mu_4)$. The corresponding mull and alternative hypotheses are $H_{01}: \mu_1 = \mu_2$, $H_{a1}: \mu_1 > \mu_2$; $H_{02}: \mu_1 = \frac{1}{2}(\mu_2 + \mu_4)$, $H_{a2}: \mu_1 > \frac{1}{2}(\mu_2 + \mu_4)$; and $H_{03}: \mu_3 = \frac{1}{3}(\mu_1 + \mu_2 + \mu_4)$, $H_{a3}: \mu_3 > \frac{1}{3}(\mu_1 + \mu_2 + \mu_4)$. **(b)** From Exercise 10.9 we find $s_p = 46.94$. The first sample contrast is

$$c = \bar{x}_1 - \bar{x}_2$$
$$= 291.91 - 308.97$$
$$= -17.06$$

with standard error

$$s_c = 46.94\sqrt{1^2/10 + (-1)^2/5}$$
$$= 25.71$$

The test statistic is $t_1 = c/s_c = -17.06/25.71 = -.66$ with $df = 32$, and $P > .25$. From the computer $P = .256$. The second sample contrast is

$$c = \bar{x}_1 - \frac{1}{2}(\bar{x}_2 + \bar{x}_4)$$

$$= 291.91 - \frac{1}{2}(308.97 + 226.07)$$
$$= 24.39$$

with standard error

$$s_c = 46.94\sqrt{1^2/10 + (-.5)^2/5 + (-.5)^2/10}$$
$$= 19.64$$

The test statistic is $t_2 = c/s_c = 24.39/19.64 = 1.24$, with $df = 32$, and $P < .15$. From the computer $P = .112$. The third sample contrast is

$$c = \bar{x}_3 - \frac{1}{3}(\bar{x}_1 + \bar{x}_2 + \bar{x}_4)$$
$$= 366.87 - \frac{1}{3}(291.91 + 308.97 + 266.07)$$
$$= 91.22$$

with standard error

$$s_c = 46.94\sqrt{(-1/3)^2/10 + (-1/3)^2/5 + 1^2/11 + (-1/3)^2/10}$$
$$= 17.27$$

The test statistic is $t_3 = c/s_c = 91.22/17.27 = 5.28$, with $df = 32$, and $P < .0005$. There is no evidence to conclude that the treatment and control groups have different means. There is no evidence to conclude that the mean of the treatment group is greater than the average of the means of the control and sedentary groups. The mean for the joggers is greater than the average of the means for the other three groups. The contrasts translate the researchers questions very well but they do not provide a complete description of the results. The treatment and control groups are quite similar; the joggers have much higher scores; and the sedentary people have much lower scores. (c) To address the question of causation in this context we would need to randomly assign some people to be sedentary and some people to be joggers. This cannot be done with human subjects. From the data we have, we cannot conclude that a sedentary lifestyle causes people to be less physically fit. Note it is possible that lack of exercise causes one to adopt a sedentary lifestyle. No conclusion about causation can be drawn.

10.18 (a) The contrasts are $\psi_1 = \mu_1 - \mu_2$, $\psi_2 = \mu_1 - \frac{1}{2}(\mu_2 + \mu_4)$, and $\psi_3 = \mu_3 - \frac{1}{3}(\mu_1 + \mu_2 + \mu_4)$. The corresponding null and alternative hypotheses are $H_{01} : \mu_1 = \mu_2$, $H_{a1} : \mu_1 > \mu_2$; $H_{02} : \mu_1 = \frac{1}{2}(\mu_2 + \mu_4)$, $H_{a2} : \mu_1 > \frac{1}{2}(\mu_2 + \mu_4)$; and $H_{03} : \mu_3 = \frac{1}{3}(\mu_1 + \mu_2 + \mu_4)$, $H_{a3} : \mu_3 > \frac{1}{3}(\mu_1 + \mu_2 + \mu_4)$. **(b)** From Exercise 10.10 we find $s_p = 7.93$. The first sample contrast is

$$\begin{aligned} c &= \bar{x}_1 - \bar{x}_2 \\ &= 51.90 - 57.40 \\ &= -5.50 \end{aligned}$$

with standard error

$$\begin{aligned} s_c &= 7.93\sqrt{1^2/10 + (-1)^2/5} \\ &= 4.3434 \end{aligned}$$

The test statistic is $t_1 = c/s_c = -5.50/4.3434 = -1.27$, with $df = 32$, and $.10 \leq P \leq .15$. From the computer $P = .107$. The second sample contrast is

$$\begin{aligned} c &= \bar{x}_1 - \frac{1}{2}(\bar{x}_2 + \bar{x}_4) \\ &= 51.90 - \frac{1}{2}(57.40 + 58.20) \\ &= -5.90 \end{aligned}$$

with standard error

$$\begin{aligned} s_c &= 7.93\sqrt{1^2/10 + (-.5)^2/5 + (-.5)^2/10} \\ &= 3.3173 \end{aligned}$$

The test statistic is $t_2 = c/s_c = -5.90/3.3173 = -1.78$, with $df = 32$, and $.025 \leq P \leq .050$. From the computer $P = .042$. The third sample contrast is

$$c = \bar{x}_3 - \frac{1}{3}(\bar{x}_1 + \bar{x}_2 + \bar{x}_4)$$
$$= 49.73 - \frac{1}{3}(51.90 + 57.40 + 58.20)$$
$$= -6.10$$

with standard error

$$s_c = 7.93\sqrt{(-1/3)^2/10 + (-1/3)^2/5 + 1^2/11 + (-1/3)^2/10}$$
$$= 2.9174$$

The test statistic is $t_3 = c/s_c = -6.10/2.9174 = -2.09$, with $df = 32$, and $.02 \leq P \leq .025$. From the computer $P = .022$. Note that Table E has entries for 30 and 40 degrees of freedom. The P-value bounds given are the same for either choice. There is no evidence to conclude that the treatment and control groups have different means. The small sample size in the control group results in very little power for this comparison. There is evidence to conclude that the mean of the treatment group is better than the average of the means of the control and sedentary groups. The mean for the joggers is better than the average of the means for the other three groups. The contrasts translate the researchers questions very well but they do not provide a complete description of the results. It appears that the four groups form two clusters. The treatment group and the joggers have similar low depression scores while the control and sedentary groups have similar high scores. (c) To address the question of causation in this context we would need to randomly assign some people to be sedentary and some people to be joggers. This cannot be done with human subjects. From the data we have we cannot conclude that a sedentary lifestyle causes people to be more depressed. Note it is possible that depression causes one to adopt a sedentary lifestyle. No conclusion about causation can be drawn.

10.19 From Exercise 10.11 we find MSE $= s_p^2 = 3.898$. Therefore, $s_p = \sqrt{3.898} = 1.974$.

$$t_{ij} = (\bar{x}_i - \bar{x}_j)/s_p\sqrt{(1/n_i) + (1/n_j)}$$
$$t_{12} = (3.7 - 3.1)/1.974\sqrt{(1/46) + (1/111)}$$

CHAPTER 10

$$= 1.73$$
$$t_{13} = (3.7 - 2.9)/1.974\sqrt{(1/46) + (1/52)}$$
$$= 2.00$$
$$t_{23} = (3.1 - 2.9)/1.974\sqrt{(1/111) + (1/52)}$$
$$= .60$$

Since the absolute values of these t statistics do not exceed $t^* = 2.41$, no pair of means is distinguishable by this method.

10.20 From Exercise 10.12 we find MSE $= s_p^2 = 72412$. Therefore, $s_p = \sqrt{72412} = 269$.

$$t_{ij} = (\bar{x}_i - \bar{x}_j)/s_p\sqrt{(1/n_i) + (1/n_j)}$$
$$t_{12} = (1217 - 844)/269\sqrt{(1/88) + (1/91)}$$
$$= 9.27$$
$$t_{13} = (1217 - 1119)/269\sqrt{(1/88) + (1/54)}$$
$$= 2.11$$
$$t_{23} = (844 - 1119)/269\sqrt{(1/91) + (1/54)}$$
$$= -5.95$$

Since the absolute values of t_{12} and t_{23} exceed $t^* = 2.41$, we conclude that the Egypt and Kenya means differ and the Egypt and Mexico means differ. The means for Kenya and Mexico are not distinguishable by this method.

10.21 From Exercise 10.9 we find $s_p = 46.94$.

$$t_{ij} = (\bar{x}_i - \bar{x}_j)/s_p\sqrt{(1/n_i) + (1/n_j)}$$
$$t_{12} = (291.91 - 308.97)/46.94\sqrt{(1/10) + (1/5)}$$
$$= -.66$$
$$t_{13} = (291.91 - 366.87)/46.94\sqrt{(1/10) + (1/11)}$$
$$= -3.65$$
$$t_{14} = (291.91 - 226.07)/46.94\sqrt{(1/10) + (1/10)}$$
$$= 3.14$$

$$t_{23} = (308.97 - 366.87)/46.94\sqrt{(1/5) + (1/11)}$$
$$= -2.29$$
$$t_{24} = (308.97 - 226.07)/46.94\sqrt{(1/5) + (1/10)}$$
$$= 3.22$$
$$t_{34} = (366.87 - 226.07)/46.94\sqrt{(1/11) + (1/10)}$$
$$= 6.86$$

Since the absolute values of t_{13}, t_{14}, t_{24} and t_{34} exceed $t^* = 2.53$, we conclude that the following pairs of groups differ: treatment and joggers, treatment and sedentary, control and sedentary, and joggers and sedentary. The means for treatment and control, and control and joggers are not distinguishable by this method.

10.22 From Exercise 10.10 we find $s_p = 7.93$.

$$t_{ij} = (\bar{x}_i - \bar{x}_j)/s_p\sqrt{(1/n_i) + (1/n_j)}$$
$$t_{12} = (51.90 - 57.40)/7.93\sqrt{(1/10) + (1/5)}$$
$$= -1.27$$
$$t_{13} = (51.90 - 49.73)/7.93\sqrt{(1/10) + (1/11)}$$
$$= .63$$
$$t_{14} = (51.90 - 58.20)/7.93\sqrt{(1/10) + (1/10)}$$
$$= -1.78$$
$$t_{23} = (57.40 - 49.73)/7.93\sqrt{(1/5) + (1/11)}$$
$$= 1.79$$
$$t_{24} = (57.40 - 58.20)/7.93\sqrt{(1/5) + (1/10)}$$
$$= -.18$$
$$t_{34} = (49.73 - 58.20)/7.93\sqrt{(1/11) + (1/10)}$$
$$= -2.44$$

None of the pairs of means are distinguishable by this method.

10.23 First, $\bar{\mu} = (\mu_1 + \mu_2 + \mu_3)/3 = (2.5 + 3.0 + 3.5)/3 = 3.0$. The noncentrality parameter is $\lambda = n((2.5-3.0)^2 + (3.0-3.0)^2 + (3.5-3.0)^2)/2.3^2 = .09452n$.

CHAPTER 10   331

n	DFG	DFE	F^*	λ	Power
50	2	147	3.06	4.73	.47
100	2	297	3.03	9.45	.79
150	2	447	3.02	14.18	.93
175	2	522	3.01	16.54	.96
200	2	597	3.01	18.90	.98

A sample size of 175 appears to be adequate.

10.24 First, $\bar{\mu} = (\mu_1+\mu_2+\mu_3)/3 = (2.7+3.0+3.3)/3 = 3.0$. The noncentrality parameter is $\lambda = n((2.7-3.0)^2+(3.0-3.0)^2+(3.3-3.0)^2)/2.3^2 = .03403n$.

n	DFG	DFE	F^*	λ	Power
50	2	147	3.06	1.70	.19
100	2	297	3.03	3.40	.36
150	2	447	3.02	5.10	.51
175	2	522	3.01	5.95	.58
200	2	597	3.01	6.81	.64
300	2	897	3.01	10.21	.82
400	2	1197	3.00	13.61	.92
500	2	1497	3.00	17.01	.97

Of the choices given in the exercise the largest sample size of 200 is preferred. Note that the power is only 64% for this choice. Further calculations indicate that a sample size of 500 or more is needed for high power.

10.25 (a)

	$\bar{x}$	s
0	10.65	2.05
1,000	10.43	1.49
5,000	5.60	1.24
10,000	5.45	1.77

(b) The hypotheses are $H_0 : \mu_1 = \mu_2 = \mu_3 = \mu_4$ and H_a : Not all of the μ_i are equal. The ANOVA tests whether or not the number of nematodes on the plants affects the growth of the seedlings. (c) $F = 12.08$, $P = .0006$. There is evidence to conclude that not all of the means are equal. $s_p = 1.67$,

$R^2 = .75$.

10.26 (a)

	$\bar{x}$	s
Lemon Yellow	47.17	6.79
White	15.67	3.33
Green	31.50	9.91
Blue	14.83	5.34

(b) The hypotheses are $H_0 : \mu_1 = \mu_2 = \mu_3 = \mu_4$, H_a : Not all of the μ_i are equal. **(c)** $F = 30.55$, $P = .0001$. There is evidence to conclude that not all of the means are equal. $s_p = 6.78$, $R^2 = .82$.

10.27 (a) $\psi = \mu_1 - \frac{1}{3}(\mu_2 + \mu_3 + \mu_4)$. **(b)** $H_0 : \mu_1 = \frac{1}{3}(\mu_2 + \mu_3 + \mu_4)$. $H_a : \mu_1 > \frac{1}{3}(\mu_2 + \mu_3 + \mu_4)$. **(c)** From computer software we obtain the following results. $t = 3.63$, df=12, $P = .002$. Yes, H_0 is rejected. **(d)** $\psi = \mu_1 - \mu_4$.

$$c = \bar{x}_1 - \bar{x}_4$$
$$= 10.65 - 5.45$$
$$= 5.2$$

$$s_c = 1.6665\sqrt{1^2/4 + (-1)^2/4}$$
$$= 1.178$$

Using Table E with 12 degrees of freedom we find $t^* = 2.179$ for a 95% confidence interval. The interval is $c \pm t^* s_c = 5.2 \pm 2.179(1.178) = (2.63, 7.77)$.

10.28 From Exercise 10.26 we find $s_p = 6.7842$.

$$t_{ij} = (\bar{x}_i - \bar{x}_j)/s_p\sqrt{(1/n_i) + (1/n_j)}$$
$$t_{12} = (47.17 - 15.67)/6.7842\sqrt{(1/6) + (1/6)}$$
$$= 8.04$$
$$t_{13} = (47.17 - 31.50)/6.7842\sqrt{(1/6) + (1/6)}$$
$$= 4.00$$

CHAPTER 10

$$t_{14} = (47.17 - 14.83)/6.7842\sqrt{(1/6) + (1/6)}$$
$$= 8.26$$
$$t_{23} = (15.67 - 31.50)/6.7842\sqrt{(1/6) + (1/6)}$$
$$= -4.04$$
$$t_{24} = (15.67 - 14.83)/6.7842\sqrt{(1/6) + (1/6)}$$
$$= .21$$
$$t_{34} = (31.50 - 14.83)/6.7842\sqrt{(1/6) + (1/6)}$$
$$= 4.26$$

All of the pairs of means are significantly different except white and blue.

10.29 (a) The first two forms have similar distributions but students who took the third form tended to have higher scores. The IQRs are similar and so are the standard deviations. The distributions appear to be symmetric. There are some low scores (more than 1.5IQR below the lower quartile for forms 1 and 2).

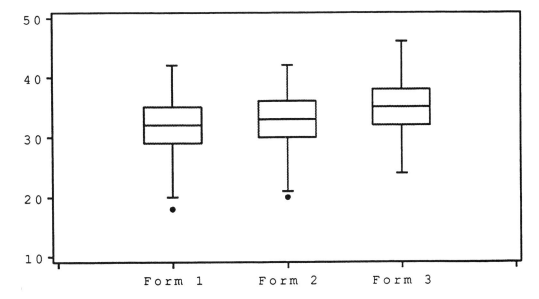

(b) The ANOVA comparing the scores for the three forms indicates that the mean scores are significantly different ($F = 7.61$; df=2,238; $P = .0006$). The Bonferroni multiple comparison procedure run at the .05 level indicates that form 3 produces higher scores than form 1. The difference in means is 2.7 points and the difference is between 1.0 and 4.4 points (with 95% simultaneous confidence). This procedure fails to detect the difference between forms 3 and 2. The difference between the means is 1.6 with a confidence interval of $-.07$ to 3.3.

10.30 (a) The concentration of lead appears to be decreasing over time. (We get the same conclusion whether we look at medians or means). The distributions are approximately symmetric. There may be a few low outliers.

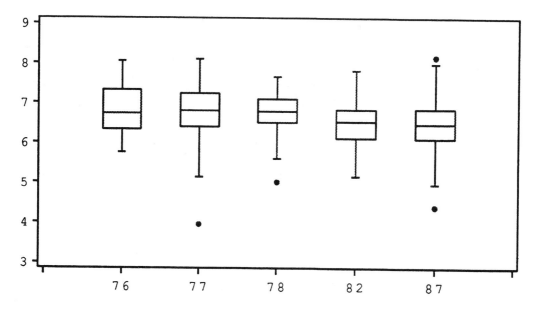

(b) An analysis of variance indicates that the lead concentrations for the five years are not all the same ($F = 5.75$; df=4,308; $P = .0002$). The Bonferroni procedure indicates that lead concentration in 1987 is less that in 1976, 1977, and 1978. No other differences between years are declared to be significant. This analysis clearly establishes the fact that the lead concentration in the Hubbard Forest floor is decreasing.

10.31 (a) The response variable is yield of tomatoes in pounds. The factors are variety with four levels, and fertilizer with two levels. The total number of observations is 40. **(b)** The response variable is attractiveness rating. The factors are type of packaging with six levels, and parts of country with four levels. The total number of observations is 720. **(c)** The response variable is weight loss. The factors are type of weight loss program with three levels, and sex with two levels. The total number of observations is 30.

10.32 (a) The response variable is number of hours of sleep. The factors are amount of smoking with three levels, and sex with two levels. The total number of observations is 600. **(b)** The response variable is strength of the cement. The factors are type of mixture with five levels, and cycles of freezing and thawing with three levels. The total number of observations is 30. **(c)** The response variable is score on the final exam. The factors are method of teaching sign language with four levels, and major with two levels. The total number of observations is 40.

10.33 (a)

Source	DF
Variety	3
Fertilizer	1
Variety x Fertilizer	3
Error	32
Total	39

(b)

Source	DF
Packaging	5
Part	3
Packaging x Part	15
Error	696
Total	719

(c)

Source	DF
Program	2
Sex	1
Program x Sex	2
Error	24
Total	29

10.34 (a)

Source	DF
Smoking	2
Sex	1
Smoking x Sex	2
Error	594
Total	599

(b)

Source	DF
Mixtures	4
Cycles	2
Mixtures x Cycles	8
Error	15
Total	29

(c)

Source	DF
Methods	3
Major	1
Methods x Major	3
Error	32
Total	39

10.35 (a)

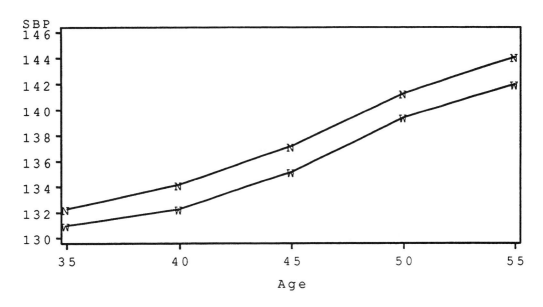

(b) Blood pressure for nonwhites is higher than for whites in all age groups. Yes, blood pressure increases with age. There does not appear to be an interaction. (c) The marginal means for race are 135.98 and 137.82. The marginal means for age groups are 131.65, 133.25, 136.20, 140.35 and 143.05. The differences between the white and nonwhite blood pressures are 1.3, 1.9, 2.0, 1.9 and 2.1.

10.36 (a)

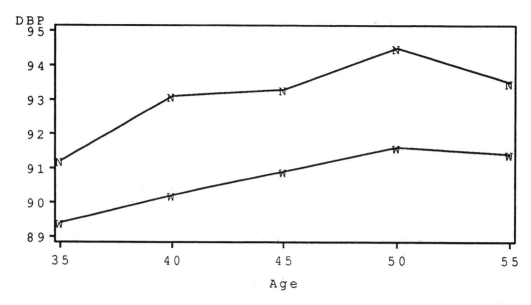

(b) Blood pressure for nonwhites is higher than for whites in all age groups. Yes, blood pressure increases with age until 50-54; then there is a small decrease. There does not appear to be an interaction. (c) The marginal means for race are 90.70 and 93.12. The marginal means for age groups are 90.30, 91.65, 92.10, 93.05 and 92.45. The differences between the white and nonwhite blood pressures are 1.8, 2.9, 2.4, 2.9 and 2.1.

10.37 (b) The mean amount of GITH for the low chromium diet is about the same as the amount for the normal chromium diet. The GITH for the R diet is greater than the M diet. Yes, there is an increase in GITH for the R diet as the level of chromium is increased; but for the M diet there is a decrease. (c) The marginal means for diet are 4.48 and 5.25. The marginal means for chromium are 4.86 and 4.87. The differences between the diets are .63 and .89. There is a larger difference for the normal chromium level.

10.38 The means for males and females are approximately equal. The psychology graduate students have a higher mean than the liberal arts undergraduates. Among the psychology graduate students the females score higher than the males, whereas for the liberal arts undergraduates this pattern is reversed. Therefore, there is an interaction. The marginal means for sex are 26.45 and 27.10. The marginal means for group are 28.41 and 25.14. The differences between the two groups of students are 2.22 and 4.31.

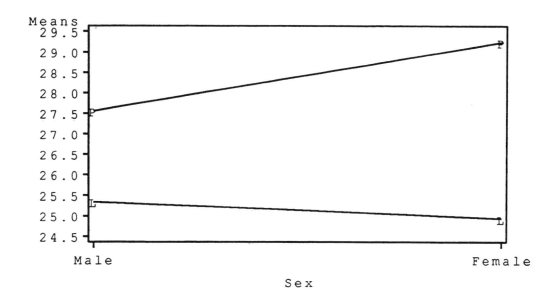

10.39 The average score for the males is higher than for the females. The scores in engineering are higher than the scores in computer science. The other group has the lowest scores. It looks like there may be an interaction. The marginal means for sex are 611.7 and 585.3. The marginal means for majors are 605.0, 624.5 and 566.0. The differences between males and females are 46, −13 and 46.

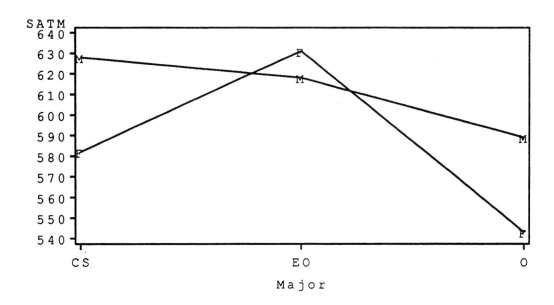

10.40 The females have higher high school math grades on the average than the males. The grades of the CS and EO majors are approximately equal and are higher than the grades of the O majors. The differences between males and females for CS and O are approximately equal. For EO there is a larger sex difference. This suggests that there is an interaction. The marginal means for sex are 8.23 and 8.84. The marginal means for majors are 8.89, 8.86 and 7.84. The differences between males and females are .43, 1.01 and .39.

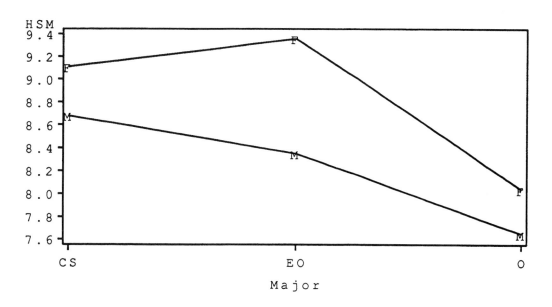

10.41 (a)

Source	SS	DF	MS	F
A	0.00121	1	0.00121	.04
B	5.79121	1	5.79121	192.89
A x B	0.17161	1	0.17161	5.72
Error	1.08084	36	0.03002	
Total	7.04487	39		

(b) For the interaction the test statistic is $F = 5.72$. It has the $F(1, 36)$ distribution. In Table F we can find entries for (1,30) and (1,40) degrees of freedom. These both give $.01 \leq P \leq .025$. The exact value is $P = .022$.
(c) For the main effect of chromium (A) the test statistic is $F = .04$. It has the $F(1, 36)$ distribution. In Table F we can find entries for (1,30) and (1,40) degrees of freedom. These both give $P > .10$. The exact value is $P = .84$. For the main effect of eat (B) the test statistic is $F = 192.89$. It has the $F(1, 36)$ distribution. In Table F we can find entries for (1,30) and (1,40) degrees of freedom. These both give $P < .001$. (d) $s_p^2 = $ MSE .03. $s_p = \sqrt{.03} = .17$. (e) The interpretation given in Exercise 10.37 is supported by the significance tests.

10.42 (a)

Source	SS	DF	MS	F
A	62.40	1	62.40	1.35
B	1599.03	1	1599.03	34.72
A x B	163.80	1	163.80	3.56
Error	13633.29	296	46.06	
Total	15458.52	299		

(b) For the interaction the test statistic is $F = 3.56$. It has the $F(1, 296)$ distribution. In Table F we can find entries for (1,200) and (1,1000) degrees of freedom. These both give $.05 \leq P \leq .10$. The exact value is $P = .060$. **(c)** For the main effect of sex (A) the test statistic is $F = 1.35$. It has the $F(1, 296)$ distribution. In Table F we can find entries for (1,200) and (1,1000) degrees of freedom. These both give $P > .10$. The exact value is $P = .245$. For the main effect of group (B) the test statistic is $F = 34.72$. It has the $F(1, 296)$ distribution. In Table F we can find entries for (1,200) and (1,1000) degrees of freedom. These both give $P < .001$. The exact value is $P = .00000001$. **(d)** $s_p^2 = \text{MSE} = 46.06$. $s_p = \sqrt{46.06} = 6.8$. **(e)** The interaction effect just fails to be statistically significant at the traditional .05 level. The group effect is significant supporting the patterns described in Exercise 10.38. No evidence of a sex effect is present.

10.43 (a) For each F statistic the degrees of freedom are 1 and 945. Both main effects are highly significant ($P < .001$) while the interaction in not ($P > .1$). **(b)** Women live longer than men and right-handed people live longer than left-handed people.

10.44 (a) The ANOVA indicates that there is a main effect for series ($F = 7.02$ with 3 and 61 degrees of freedom; $P = .0004$); there is no evidence of a main effect for holder ($F = 1.96$ with 1 and 61 degrees of freedom; $P = .1665$); and there is no evidence of an interaction between series and holder ($F = 1.24$ with 3 and 61 degrees of freedom; $P = .3026$). There is no evidence from this experiment to suggest that the holders make a difference in the radon readings. **(b)** The ANOVA indicates that detectors from different series give different radon measurements when exposed to the same source. The means for the four series should be examined by a radon expert to determine whether or not this kind of variation is a serious problem. In comparative studies using detectors of this type, the researchers should be

10.45 (a)

	n	$\bar{x}$	s
Basal	22	5.27	2.76
DRTA	22	5.09	2.00
Strat	22	4.95	1.86

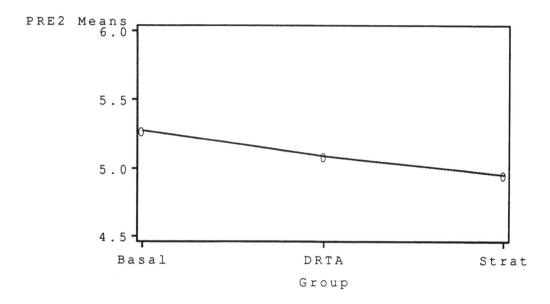

(b) The data appear to be approximately normal.

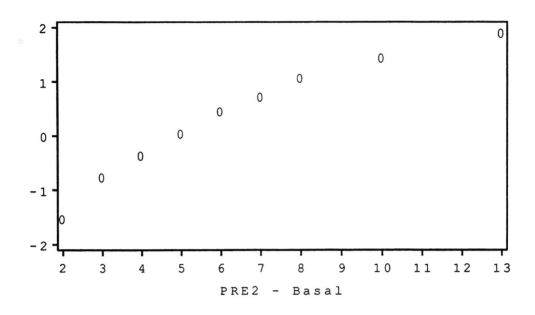
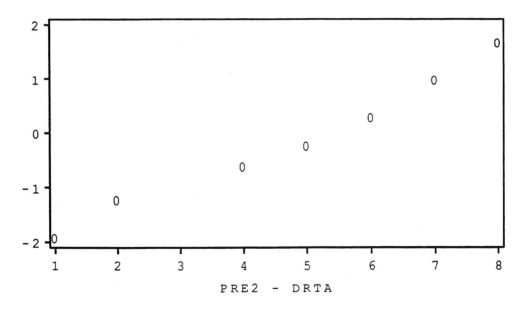

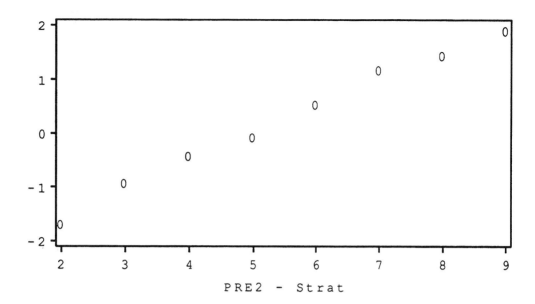

(c) The ratio of the largest to the smallest standard deviation is $2.76/1.86 = 1.48$. Since the ratio is less than two we proceed with the analysis of variance.
(d) The hypotheses are $H_0 : \mu_1 = \mu_2 = \mu_3$ and H_a : not all of the μ_i are equal. $F = .11$, $P = .8948$. H_0 is not rejected. (e) There is no evidence to conclude that the three means are different.

10.46 (a)

	n	$\bar{x}$	s
Basal	22	6.27	2.76
DRTA	22	5.09	2.00
Strat	22	3.95	1.86

(b) The test statistic is $F = 5.87$ with 2 and 63 degrees of freedom. There is evidence to conclude that the three groups are not all the same ($P = .0046$).
(c) In Exercise 10.41 we did not have evidence to conclude that the means were different. In this exercise we find that they are different. (d) By making the means further apart we have increased the value of the F statistic. This occurred because the variability of group means increased while the variability within groups remained the same.

10.47 (a)

	n	$\bar{x}$	s
Basal	22	6.68	2.77
DRTA	22	9.77	2.72
Strat	22	7.77	3.93

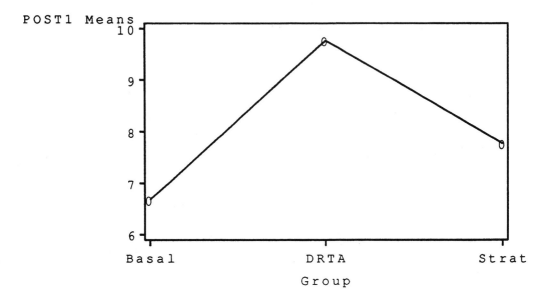

(b) The distributions look approximately normal.

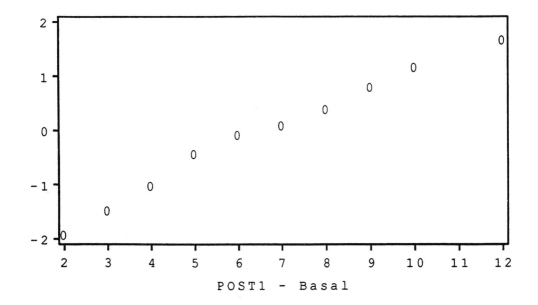

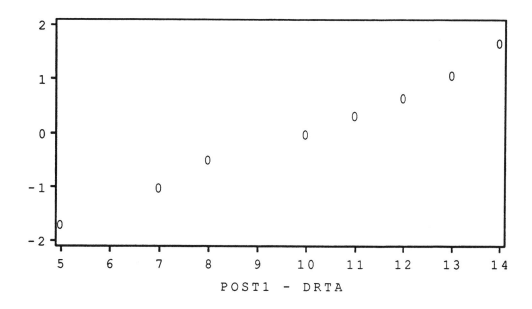

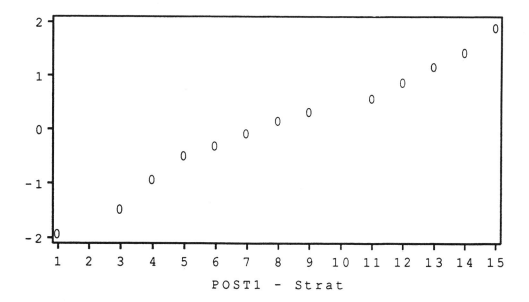

(c) The ratio of the largest to the smallest standard deviation is $3.93/2.72 = 1.44$. Since the ratio is less than two we proceed with the analysis of variance. (d) $F = 5.32$, $P = .0073$. H_0 is rejected. (e) $\psi = -\mu_1 + \frac{1}{2}(\mu_2 + \mu_3)$. $H_0 : \mu_1 = \frac{1}{2}(\mu_2 + \mu_3)$, $H_a : \mu_1 < \frac{1}{2}(\mu_2 + \mu_3)$. $c = 2.09$, $s_c = .833$, $t = 2.51$, df=63, $P = .007$. The 95% confidence interval is $(.43, 3.75)$. **(f)** $\psi = \mu_2 - \mu_3$. $H_0 : \mu_2 = \mu_3$, $H_a : \mu_2 \neq \mu_3$. $c = 2.00$, $s_c = .96$, $t = 2.08$, df=63, $P = .0415$. The 95% confidence interval is $(.079, 3.921)$. **(g)** Not all of the means are equal. The Basal group mean is smaller than the average of the DRTA and Strat group means, and the DRTA group mean is larger than the Strat group mean.

10.48 (a)

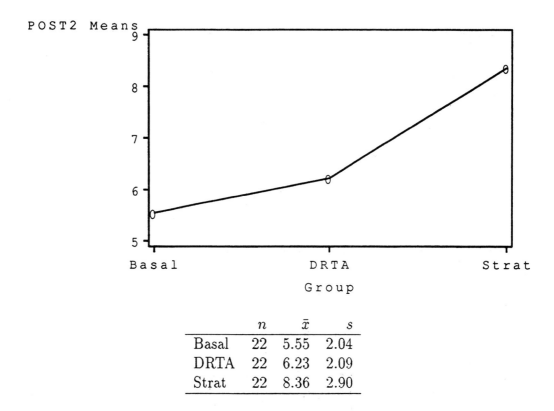

	n	$\bar{x}$	s
Basal	22	5.55	2.04
DRTA	22	6.23	2.09
Strat	22	8.36	2.90

(b) The distribution for the Basal group is skewed to the right. Slightly less than one half of the observations have a value of 6 in the DRTA group. There is one somewhat low value in the Strat group. None of these deviations from normality appear to be serious.

CHAPTER 10

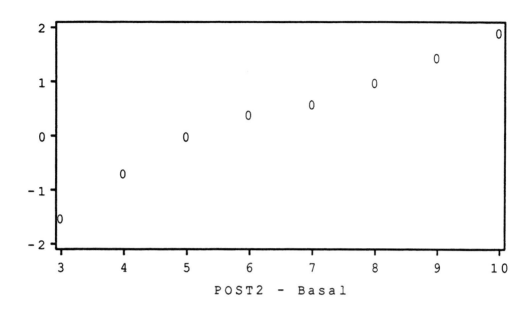

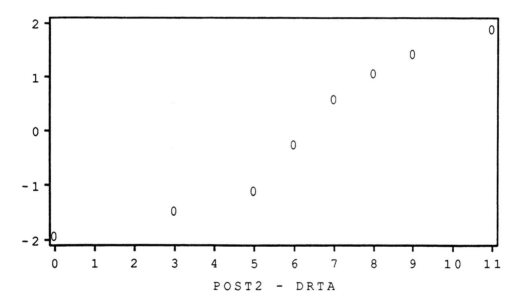

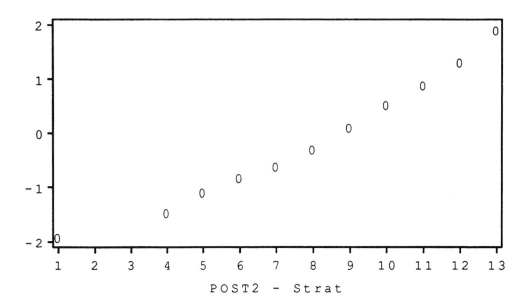

(c) The ratio of the largest to the smallest standard deviation is $2.90/2.04 = 1.42$. Since the ratio is less than two we proceed with the analysis of variance. (d) $F = 8.41$, $P = .0006$. H_0 is rejected. (e) $\psi = -\mu_1 + \frac{1}{2}(\mu_2 + \mu_3)$. $H_0 : \mu_1 = \frac{1}{2}(\mu_2 + \mu_3)$, $H_a : \mu_1 < \frac{1}{2}(\mu_2 + \mu_3)$. $c = 1.745$, $s_c = .6211$, $t = 2.81$, df=63, $P = .0033$. The 95% confidence interval is $(.50, 2.99)$. (f) $\psi = \mu_2 - \mu_3$. $H_0 : \mu_2 = \mu_3$, $H_a : \mu_2 \neq \mu_3$. $c = -2.13$, $s_c = .7171$, $t = -2.97$, df=63, $P = .0042$. The 95% confidence interval is $(-3.56, -.70)$. (g) Not all of the means are equal. The Basal group mean is smaller than the average of the DRTA and Strat group means, and the Strat group mean is larger than the DRTA group mean.

10.49 (a) $F = 1.33$, $P = .31$. H_0 is not rejected. (b) $F = 12.08$, $P = .0006$, from Exercise 10.25. The outlier has caused a statistically significant result to disappear. (c)

	$\bar{x}$	s
0	34.95	48.74
1,000	10.43	1.49
5,000	5.60	1.24
10,000	5.45	1.77

The large standard deviation and mean for the first group relative to the other groups should lead us to check the data in this group.

10.50 (a) $F = 2.00$, $P = .146$. H_0 is not rejected. **(b)** $F = 30.55$, $P = .0001$, from Exercise 10.26. The outlier has caused a statistically significant result to disappear. **(c)**

	$\bar{x}$	s
Lemon Yellow	114.67	164.42
White	15.67	3.33
Green	31.50	9.91
Blue	14.83	5.34

The large standard deviation and mean for the first group relative to the other groups should lead us to check the data in this group.

10.51 (a)

	$\bar{x}$	s
0	1.02	.08
1,000	1.01	.07
5,000	.74	.09
10,000	.72	.15

(b) The hypotheses are $H_0 : \mu_1 = \mu_2 = \mu_3 = \mu_4$ and H_a : Not all of the μ_i are equal. The ANOVA tests whether or not the number of nematodes on the plants affects the growth of the seedlings. **(c)** $F = 10.39$, $P = .0012$. The number of nematodes affects the growth of the seedlings. $s_p = .10$, $R^2 = .72$. The results are qualitatively the same.

10.52 (a)

	$\bar{x}$	s
Lemon Yellow	6.85	.49
White	3.94	.41
Green	5.54	.96
Blue	3.79	.73

(b) The hypotheses are $H_0 : \mu_1 = \mu_2 = \mu_3 = \mu_4$ and H_a : Not all of the μ_i are equal. **(c)** $F = 27.00$, $P = .0001$. There is evidence to conclude that not all of the means are equal. $s_p = .6835$, $R^2 = .80$. The results are essentially the same as those given in Exercise 10.26.

10.53 (a) First, $\bar{\mu} = (\mu_1+\mu_2+\mu_3+\mu_4)/4 = (620+600+580+560)/4 = 590$. The noncentrality parameter is $\lambda = n((620-590)^2 + (600-590)^2 + (580-590)^2 + (560-590)^2)/90^2 = .2469n$.

n	Power
20	.42
40	.74
60	.91
80	.97
100	.99

(b) The power increases rapidly up to 60 and levels off afterward.

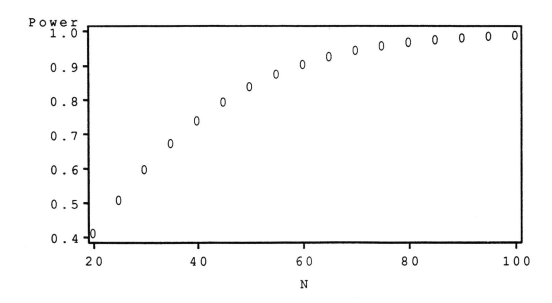

(c) A sample size between 80 and 100 will give good power.

10.54 (a) First, $\bar{\mu} = (\mu_1+\mu_2+\mu_3+\mu_4)/4 = (610+600+590+580)/4 = 595$. The noncentrality parameter is $\lambda = n((610-595)^2 + (600-595)^2 + (590-595)^2 + (580-595)^2)/90^2 = .6173n$.

CHAPTER 10   353

n	Power
50	.28
100	.53
200	.85
300	.96
400	.99

(b) The power increases rapidly up to 200 and levels off afterward.

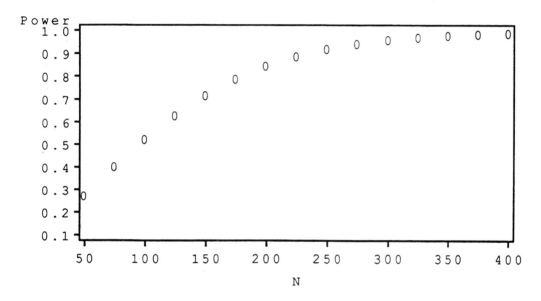

(c) A sample size between 300 and 400 will give good power.

10.55 (a)

Species	Size	$\bar{x}$	s
Aspen	S1	387	69
Aspen	S2	336	60
Birch	S1	293	122
Birch	S2	455	118
Maple	S1	323	52
Maple	S2	294	184

The marginal means for species are 362, 374 and 308. The marginal means for size are 334 and 362. The table of means and marginal means is

	S1	S2	
Aspen	387	336	362
Birch	293	455	374
Maple	323	294	308
	334	362	

(b) Both the species and the sizes appear to be different. The interaction is so strong that statements about the marginal means are meaningless.

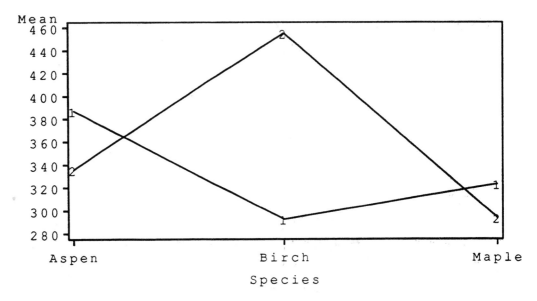

(c) No statistically significant effects are present. The impressions obtained from the plot are not substantiated by the analysis and are probably due to chance variation. Note that there is large range of standard deviations and the sample sizes are small.

10.56 (a)

Species	Size	$\bar{x}$	s
Aspen	S1	1660	351
Aspen	S2	1321	235
Birch	S1	1168	236
Birch	S2	1345	96
Maple	S1	1526	308
Maple	S2	1142	356

The marginal means for species are 1491, 1257 and 1334. The marginal means for size are 1451 and 1270. The table of means and marginal means is

	S1	S2	
Aspen	1660	1321	1491
Birch	1168	1345	1257
Maple	1526	1142	1334
	1451	1270	

(b) Both the species and the sizes appear to be different. The interaction is so strong that statements about the marginal means are meaningless.

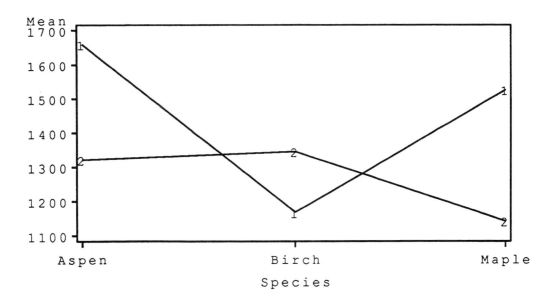

(c) No statistically significant effects are present. The impressions obtained from the plot are not substantiated by the analysis and are probably due to chance variation. Note that there is large range of standard deviations and the sample sizes are small.

10.57

Sex	Major	n	$\bar{x}$	s
Male	CS	39	526.95	100.94
Male	EO	39	507.85	57.21
Male	O	39	487.56	108.78
Female	CS	39	543.38	77.65
Female	EO	39	538.21	102.21
Female	O	39	465.03	82.18

The normal quantile plots are

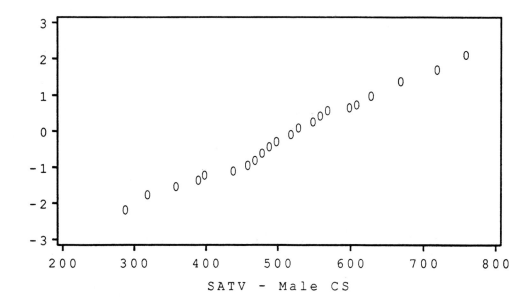

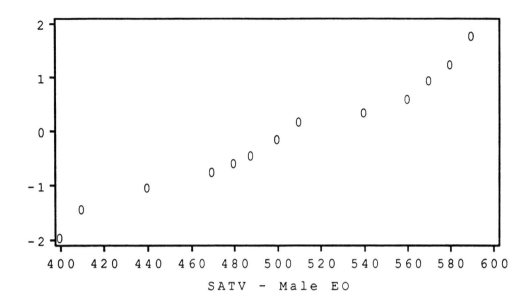

SATV - Male EO

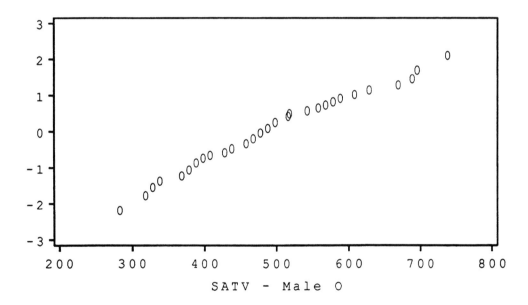

SATV - Male O

358 SOLUTIONS TO EXERCISES

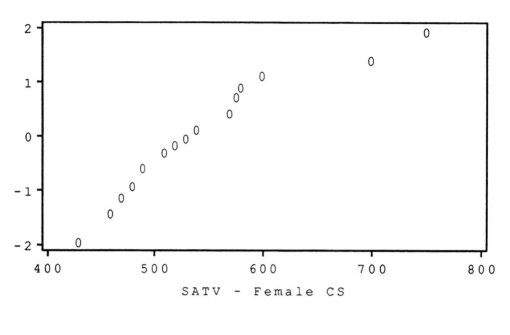

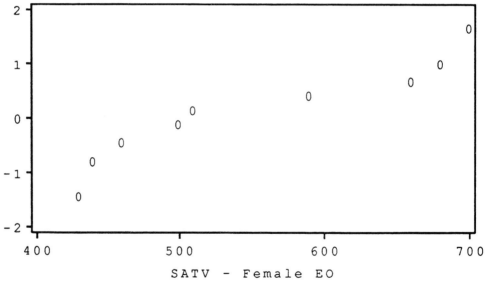

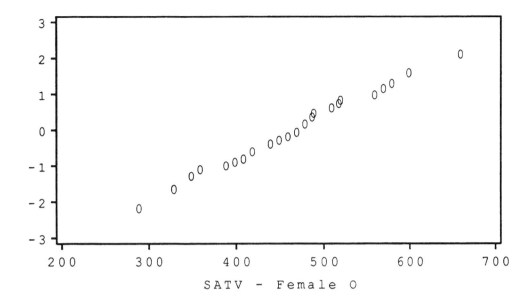

The following is a plot of the means

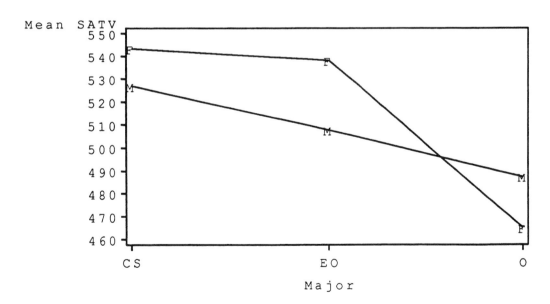

The interaction is not significant ($F = 1.81$, $P = .17$). The sex effect is not significant ($F = .47$, $P = .49$). The majors differ significantly ($F = 9.32$, $P = .0001$).

10.58

Sex	Major	n	$\bar{x}$	s
Male	CS	39	8.67	1.28
Male	EO	39	7.92	2.06
Male	O	39	7.44	1.71
Female	CS	39	8.38	1.66
Female	EO	39	9.23	0.71
Female	O	39	7.82	1.80

The normal quantile plots are

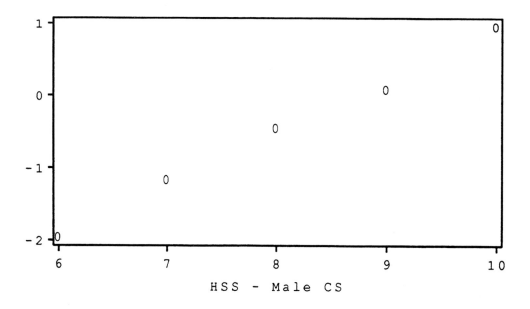

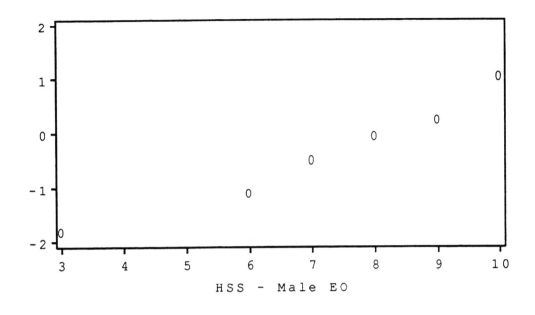

HSS - Male EO

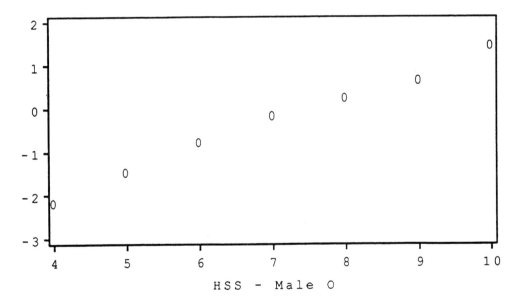

HSS - Male O

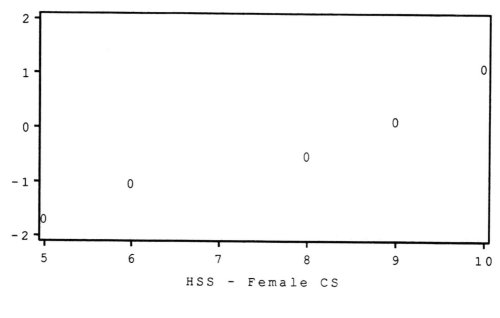

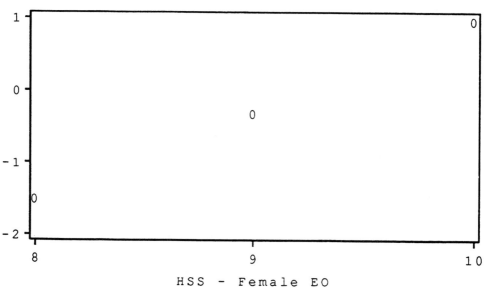

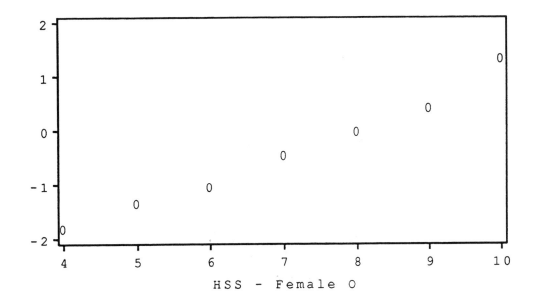

The high school grade variables are clearly not normal. We rely upon the robustness of the ANOVA procedures for this analysis. The following is a plot of the means

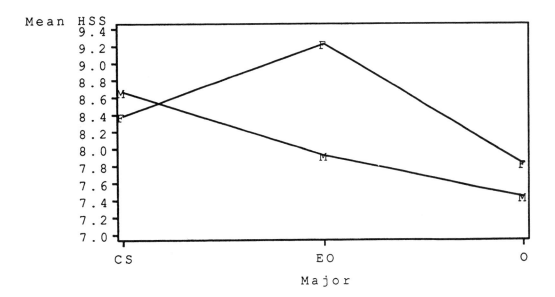

The interaction is significant ($F = 4.86$, $P = .0086$). The sex effect is significant ($F = 5.06$, $P = .0255$). The majors differ significantly ($F = 8.69$,

$P = .0002$).

10.59

Sex	Major	n	$\bar{x}$	s
Male	CS	39	7.79	1.51
Male	EO	39	7.49	2.15
Male	O	39	7.41	1.57
Female	CS	39	8.85	1.14
Female	EO	39	9.26	.75
Female	O	39	8.62	1.16

The normal quantile plots are

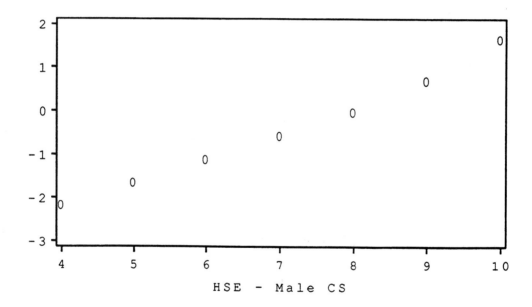

CHAPTER 10

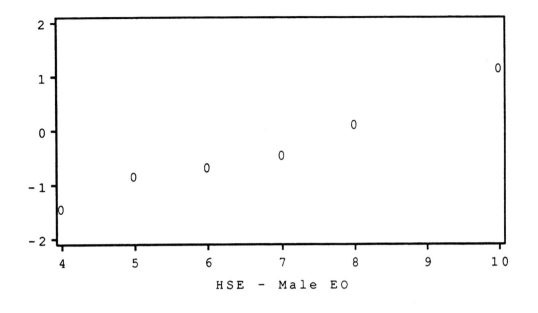

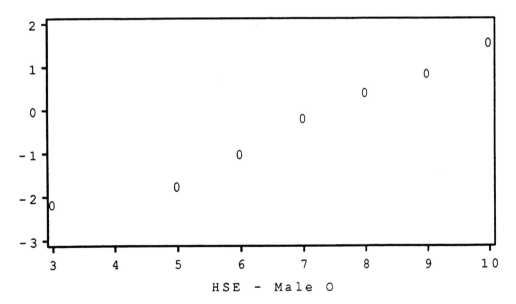

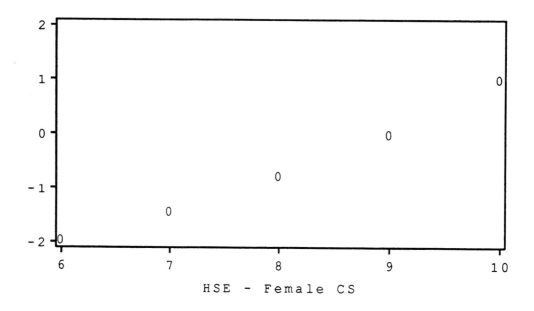

HSE - Female CS

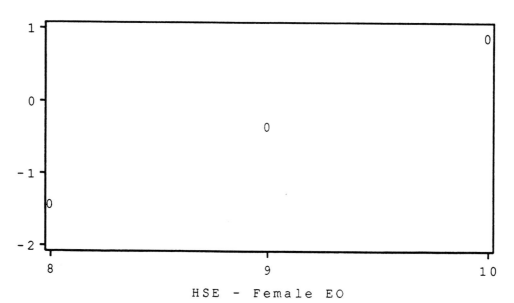

HSE - Female EO

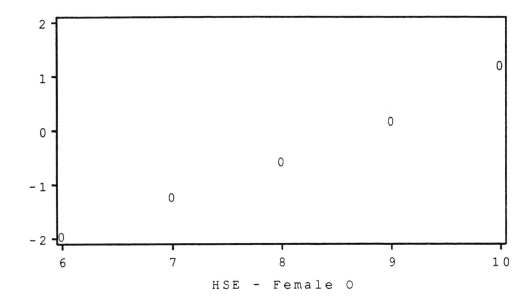

The high school grade variables are clearly not normal. We rely upon the robustness of the ANOVA procedures for nonnormal data for this analysis. The following is a plot of the means

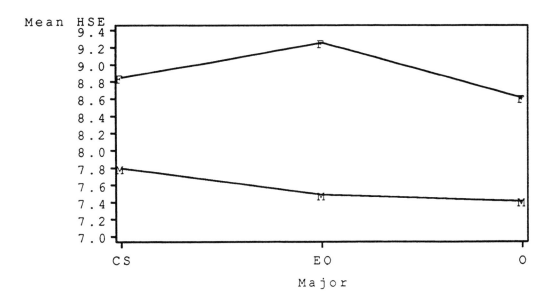

The interaction is not significant ($F = 1.33$, $P = .27$). There is a significant sex difference ($F = 50.32$, $P = .0001$). The majors are not significantly

different ($F = 1.40$, $P = .25$).

10.60

Sex	Major	n	$\bar{x}$	s
Male	CS	39	4.75	.684
Male	EO	39	5.10	.513
Male	O	39	4.05	.730
Female	CS	39	4.98	.533
Female	EO	39	5.08	.648
Female	O	39	4.52	.766

The normal quantile plots are

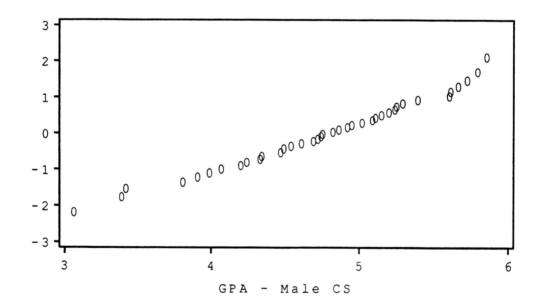

CHAPTER 10

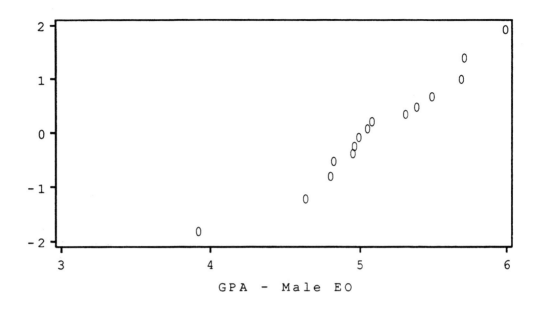

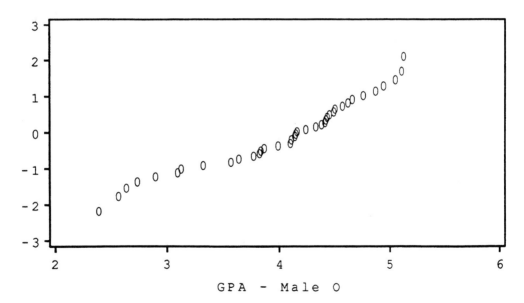

370 SOLUTIONS TO EXERCISES

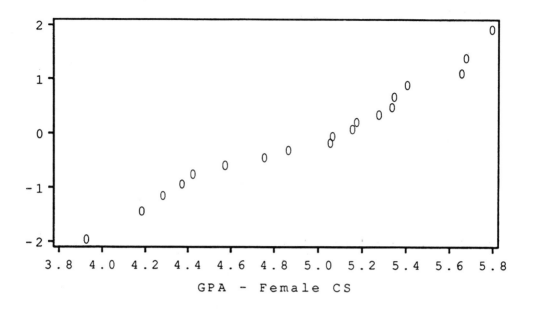

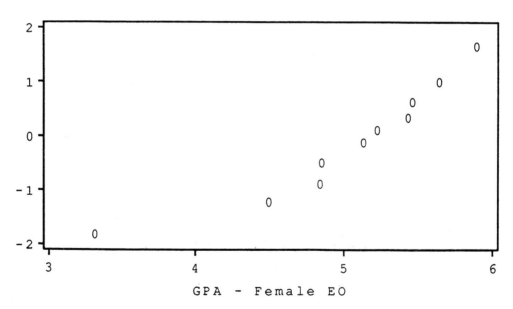

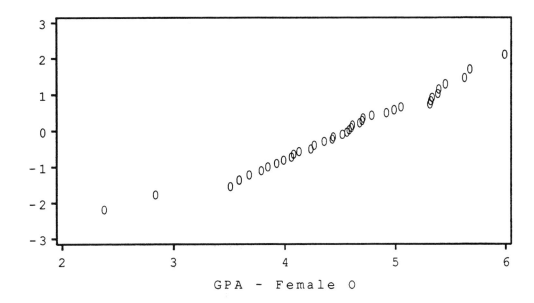

The following is a plot of the means

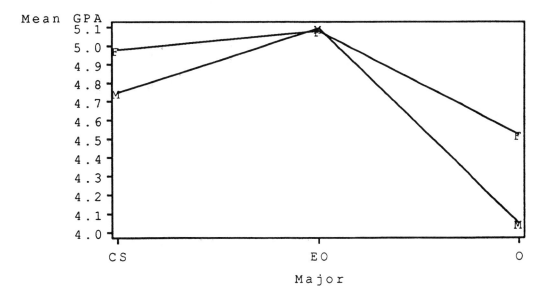

The interaction is not significant ($F = 2.77$, $P = .065$). The sex effect is significant ($F = 7.31$, $P = .007$). The majors differ significantly ($F = 31.42$, $P = .0001$).